writing
DOWN
YOUR
SOUL

How to Activate
and Listen to the
Extraordinary Voice Within

Janet Conner

Conari Press

First published in 2008 by Conari Press,
an imprint of Red Wheel/Weiser, LLC
With offices at:
500 Third Street, Suite 230
San Francisco, CA 94107
www.redwheelweiser.com

ISBN: 978-1-57324-356-8
Library of Congress Cataloging-in-Publication Data available upon request.

Cover and text design by Donna Linden
Typeset in Garamond and Art Craft
Cover photograph © Nic Taylor/iStockphoto

Printed in Canada
TCP
10 9 8 7 6 5 4 3 2

For you. Welcome to the conversation.

The Spirit's Hands
They
can be a great help— words.
They can become the spirit's hands
and lift and
caress
you.
MEISTER ECKHART

Contents

Before We Begin

THERE IS A VOICE INSIDE YOU. There is a Voice inside everyone. Whether you hear it or not, the Voice is there. Whether you acknowledge it or not, the Voice is there. Whether you ask it for help or ignore its guidance, the Voice is still there. Waiting. It is waiting for you to stop, if just for a moment, and listen. The Voice is always there, guiding you, encouraging you, loving you. This book is about connecting with that Voice.

I'll let you in on a sweet little secret right here on the very first page: connecting with that Voice is easy. And why shouldn't it be? The Voice isn't trying to hide from you—it is seeking you. It knows the rich conversation that awaits you both. It knows what you need and longs to give it to you. So it stays close at hand, in your heart, your mind, your soul. The Voice is right there, barely below the surface, waiting for you to pick up your pen and penetrate the thin wall of consciousness that keeps you apart.

But why the pen? Why writing? After all, there are other ways to connect. There are powerful spiritual and religious traditions like meditation, prayer, and ritual. There are rich body-mind-spirit practices such as massage, Reiki, yoga, and tai chi. For some, long-distance swimming or running are transcendental experiences. My son swears he finds the greatest peace and does his best thinking riding his motorcycle late at night when he's the only one on the road. All these things are good. And all of them work.

Nothing in *Writing Down Your Soul* is intended to supplant or alter the practices you use or the beliefs you hold. Deep soul writing doesn't replace anything; it enriches everything. Writing focuses your attention so clearly on the wisdom within that you cannot help but feel guided and loved. A young woman in a Writing Down Your Soul workshop expressed her surprise when she discovered how little effort was required to make that connection. "This is so easy," she said. "You don't have to listen to a CD or buy a program, or change your beliefs, or fix your diet, or anything. Just show up. Really that's it, just show up."

She's right. This kind of writing is easy. There's no one standing over your shoulder judging your grammar or punctuation or determining if anything you've said makes a lick of sense. But make no mistake, the practice of pouring your soul onto paper is profound, and, in the way of all things profound, it can—and will—change your life. Before you turn another page, consider this carefully: if you like your world the way it is, if you don't want to (or need to) improve your emotional, spiritual, or financial life, if you are content with your relationships, your family, your work, and your home, put this book down! Don't read another word. I mean it.

Because once you open that door in your soul, you can't quite close it again. You can't pretend that you don't know where the door is or how easy it is to walk through. Once you start engaging in rich, deep conversation with something higher, bigger, deeper, and wiser than yourself, you'll find yourself contemplating ideas you've never considered, saying things you've never said, and asking questions you've never asked. Once you open yourself to divine direction, you will receive guidance, but—fair warning—it may not be the guidance you expect. Once you start asking for more, you will start receiving more: more ideas, more intuition, more inspiration, more wisdom, more opportunities, more challenges, and more questions. Always, there are more questions. Because the answers, as you are about to discover, live deep inside the questions.

And let's not forget miracles. Ask and you shall receive. Every spiritual tradition tells us that asking and receiving is the law of the universe, and the Voice is happy to comply. Pick up a pen with the intention of connecting with that extraordinary Voice within, and your life will start rumbling, shifting, and moving. Awakening, as if from a long sleep, you will see your world differently, and you'll find yourself changing, subtly at first. Then, as your trust in the wisdom of the Voice expands, you'll find you have the inner strength and confidence to create your own brave new world.

Sound a bit scary? Well, the best ideas are. We all want safety, but safety, it turns out, is a paradox. To feel really safe, you first have to step out into the unknown, experience the fear, and discover that all is well. I can tell you for ten pages or ten hours that you are safe and loved, but until you feel it—feel it in the deepest place in your soul—you don't know it and certainly don't believe it. You have to step out into that space between here and there, between "who I am" and "who I could be," between "what I have" and "what I want." Nothing new can happen until you step into that empty space. Like Indiana Jones in *Raiders of the Lost Ark*, you have to thrust one foot forward into empty air and put it down firmly trusting that something somehow will prevent you from falling. And something will. Something will remind you to be not afraid. Something will encourage you to explore the possibilities. Something will talk you through the scary parts, and something will definitely celebrate your joys. From here on, the Voice will guide you. It will let you know that you are safe and loved.

Are you ready to begin? Then, by virtue of intention, you are now officially the writer of your soul. Welcome to the profound practice of entering your soul and recording the messages you find there. Let the conversation begin.

How I Discovered the Voice—
or Rather,
How the Voice Discovered Me

IT'S A SURPRISE TO ME and everyone I know that I'm the author of a book on deep soul writing. The truth is, I was never much of a journaler. Sure, when I was upset, I'd grab a notebook and write furiously for a day or two, but never consistently and never long enough to resolve anything. Mind you, I loved the *idea* of having a rich spiritual life. I loved to imagine myself sipping tea and writing profound thoughts in a tooled leather journal with morning sun dappling the pages. To bring this fantasy to life, I bought *The Artist's Way* by Julia Cameron, but it sat on a shelf alongside all the other great spiritual books I would read as soon as I had the time.

Meanwhile, I had a consulting career. I had clients and projects and reports. I had appointments and lunch dates and speaking engagements. I was a busy woman—a woman with no time to journal.

Until November 1, 1996.

I had caught my husband sleeping with his secretary the summer before. He moved out in September, but he didn't move on. On October 31, our Halloween-crazed seven-year-old begged me to invite his dad to join us for our annual Halloween extravaganza. But after trick or treating, my husband wouldn't leave. He thought we should have sex. When I refused, he shoved me out the door. He screamed that I'd never see my child again. He drank. He broke furniture. He cried. He drank some more. When he finally left at one in the morning, I collapsed into a dense, dark sleep. At dawn, my eyes shot open, five words rocketing to the surface: I am afraid of you. Those five words changed my life.

I called my husband at noon and told him I wanted a divorce. He didn't say much. Too hung over, I thought. At five, he called back. In a flat, barely audible voice he told me he had a shotgun in his mouth and was calling to say goodbye.

My mind raced. What do I do? All I could think of were those movies with the main character frantically trying to keep the other guy on the phone. Keep him talking. That's it—keep him talking.

I talked first. I talked about our son, our beautiful son. I asked questions. I asked how he felt, what he'd eaten, what was happening at work. He began to talk—just a few mumbled words, but he was saying something. Suddenly, in the middle of a sentence, he hung up. No goodbye. No grunt. No shot. No nothing. Terrifying headlines flashed across my mind: "Estranged Husband Kills Family," "Man Shoots Wife, Then Self."

I started calling friends. They all had perfectly reasonable explanations for why my son and I couldn't stay with them: I'd love to, but my husband doesn't think it's a good idea. We don't really have the room, you know. I don't think your son would be comfortable here, do you? Are you sure that's really necessary? Maybe he's just trying to scare you. Can't you stay with a neighbor?

Well, no, I couldn't stay with a neighbor. I had called my neighbor first, and he didn't want to "take sides." Desperate, I called another second grader's mother—a single mom I barely knew. Before I could finish, she stopped me. "Come straight here," she said, "I'll back my car out of the garage. Pull right in. Don't worry about clothes or food. I'll take care of everything." I grabbed my son and our Great Dane puppy and hustled them out the door.

My husband did not kill himself that night, but from then on, my family was pretty sure he was going to kill me. His rages often made it look like they were right. Overnight, my professional life disappeared. Clients have a hard time sticking around when you go into hiding every other month. Friends stop coming when they see you wearing a police emergency call button around your neck. So, did I start journaling? No, I did not. I sat and cried in the living room, with the phone unplugged so I wouldn't hear his threats, and the blinds down so he couldn't see me if he drove by.

My mother, like all good Catholic women of her time, loved to say, "God works in mysterious ways." Whenever something ludicrous happened, I'd say, "OK, Mom, how could *that* possibly be God's doing?" And she'd say, "Well, dear, God works in mysterious ways." I always thought that saying was a complete copout.

Until Harley, our Great Dane puppy, took things into his own hands—or rather, teeth.

I was sitting in my usual position, sniffling and dabbing my eyes, when I realized Harley was no longer resting his head on the ottoman and looking up at me with that consummate Great Dane blend of sadness and devotion. "Harley," I called, "where are you?" I could hear him in the hallway, and I got up to find him. He was loping slowly toward me, struggling to carry something too heavy for his scrawny neck. I pulled his burden out of his mouth—and laughed. It was *The Artist's Way*—now decorated with ripped corner, teeth marks, and Dane drool.

I wiped it off, sat down, and began to read. On page 15, I stopped cold:

> Anyone who faithfully writes morning pages will be led to
> a connection with a source of wisdom within. When I am
> stuck with a painful situation or problem that I don't think
> I know how to handle, I will go to the pages and ask for
> guidance.

Julia Cameron was talking to *me!* I needed wisdom, I most certainly was stuck in a painful situation, and I sure didn't know how to handle it. It was pretty clear that sitting and sobbing was not solving my problems. I hunted up a cheap black notebook in my office and an old brown fountain pen. The book said to write three morning pages. Well, it was morning, and at long last, I had all the time in the world to write.

But I didn't follow the directions—that is, not *The Artist's Way*'s directions. Something happened when I read that passage. My soul's needle, which had been careening madly around its compass for weeks, snapped to true north and picked up some silent subterranean instructions that guided me to write in a unique way.

"Dear God," I began. I have no idea why I started that way. It just felt right—necessary, actually. Whenever my parents were frightened, they threw themselves to their knees and begged God for help. I guess I was doing the same thing in my own way. Of course, they prayed rosaries. Me? I vented. Oh lord, how I vented! I fussed and fumed at God. "Are you paying *any* attention? Do you see what's happening here? Do you care? How are we going to live through this? How can I protect my baby? What am I going to do? Where *are* you?"

I didn't write three pages that morning; I wrote *thirty*. That was a clue that I had something to say and writing was somehow helping me say it. After an hour and a half of furious, full-speed-ahead scribbling, I didn't have any answers, but I did feel a little bit better, a little bit cleaner, a little bit lighter.

So the next morning I did it again. Day after day, I stabbed at the page in angry black ink. I told God every last little detail of every last little thing that was happening: What my husband did or threatened to do. How I cancelled my son's birthday party because his father said he'd show up with a gun. What happened when he broke into our house. How it felt to protect my son with my body. What happened when we called the police the first time, the second, the third, and the fourth. How the school insisted I drop my son off late and pick him up early to prevent scenes at school. How I moved from one coffee shop to another until it was time to pick him up. How I couldn't eat. How my son couldn't sleep. How he gnashed his teeth all night. How he crept into my bed and would not leave. How we startled in the dark at every creak and crack. How he crawled onto my lap and rocked silently for thirty minutes before he would leave for visitations with his father.

After a while, I noticed something. Not the first day or the second, but one day, there it was: a little bit of wisdom on the page.

Not the answer to my life's problems, but definitely guidance for the day's. Occasionally the answer was what to do or what not to do, but most of the time, it was something smaller, something subtler, and perhaps something richer: how to shift my thinking.

The first time it happened, I stopped writing and stared at the page. Huh? That wasn't my voice. I didn't write that. I'd never even had that thought before. But there it was. And I knew, somehow just knew, that this guidance was important. This guidance was it. This guidance was my salvation. So I followed that guidance. Like Hansel in the fairy tale, I didn't know where I was or where I was going, but I followed those precious crumbs of wisdom. Step by step, day by day, journal entry by journal entry, I inched forward.

Every morning I wrote, "Dear God," and every morning the Voice answered. One Saturday morning, I wrote about how powerless I felt when I suddenly realized that the newspaper article I was reading about an unsolved road-rage crime described my ex-husband and his truck perfectly—and that the crime had occurred thirty minutes after he had picked up our son the day before. The Voice wrote to me about the true nature of power. I prayed and tapped into that power and brought my son safely home without leaving my chair.

I wrote about the heartache of listening to a voicemail of my son struggling under his father's screaming command to "say it!" until his little voice squeaked, "Mom, you are a lying sack of shit." And the Voice wrote to me about size. It asked me which was bigger, this terrible thing or the divine? I knew the answer and turned my problem over to the divine.

I wrote about having an enemy—a big scary enemy. I asked the Voice what I should do about my enemy. The Voice told me to love my enemy. I didn't like that. And, I confess, I didn't do it—not for a long, long time.

I wrote about how scared and weak and helpless I felt, like a person riddled with holes. What's wrong with me? I cried. And the Voice wrote about strength—true strength.

I wrote about court. Twelve times I cried all over the pages telling the Voice that no matter what evidence I presented—the road-rage incident, the voicemail recording, four police reports, parents who testified to my ex-husband's threats, proof of guns in his house—the legal system insisted our son have regular, unsupervised visits with his father.

The Voice listened, wiped my tears, and listened some more.

I told the Voice how my son cried before visitation. "Tuesdays," he sobbed, "I hate Tuesdays, because after Tuesday comes Wednesday and on Wednesday I have to see my dad." I told the Voice to protect my baby when he was at his father's. The Voice always did.

I wrote about my ex-husband's weapons. The Voice asked about mine. "Words," I admitted, "words are my weapons." And the Voice helped me put my weapons down.

I wrote a list of all the things I didn't want to do but had to do in my marriage. The Voice talked to me about the difference between "have to" and "choose to." I wrote about how I disappeared into a secret waiting room in my heart when I couldn't bear what was happening. The Voice talked to me about the beautiful language of no.

I wrote about all the dreadful decisions I'd made and how badly they'd all turned out. The Voice talked to me about forgiving myself.

I wrote about my frustration waiting for the judge to let me move back to my family in Wisconsin. And the Voice talked about being frustrated waiting for me to become who I really am. "Help me remember," I said. "Who is this frightened woman?" And the Voice said, "Unafraid."

Unafraid—it was a lovely thought, a momentous thought, but I felt quite the opposite. Frightened, broke, and alone would be more like it.

Well, maybe not alone. After all, I was having real conversations in my soul journal with a divine Voice, so how alone could I be? And I was getting answers. And my life was slowly changing. Each morning I was a bit stronger, a bit wiser, a bit more aware that somehow I was going to be all right. A wee part of me kept raising her tiny head and proclaiming, "I'm going to heal. Not just survive. That's not good enough. I want to be whole and happy again!" I thought that little woman was nuts, but occasionally I let her have her say.

If I was ever going to make it all the way to healed and happy, I needed a miracle—quite a few actually. I was getting guidance. I was learning to shift my thinking. Couldn't I get miracles, too? I mean real miracles, things of substance—money, to be precise. So I asked. One morning, I wrote: "Dear God, you know I need ten thousand dollars for the attorney. I don't know how you're going to do it, but I know you're going to send ten thousand dollars. Thank you right here and right now for your gift of ten thousand dollars."

Nothing happened. There were no brilliant words, no lottery numbers, no treasure maps. I got up and made a pot of tea. Two days later my mother called.

"Dear," she said, "we've given money to all the other children but we've never given any to you. So, dear, we're sending you ten thousand dollars." (Her use of "we" was precious; my father, the other half of "we," had been dead for five years.) I said thank you, of course, to my mother, but I also wrote a profound deep thank you to my real source in my soul journal the next day.

The ten thousand dollars covered my legal bills, but it didn't touch the house expenses. I had enough savings to last almost a year. Spending it on the house was foolhardy, but everything else was blowing up in my son's life; I wanted him to be able to stay in the only house he'd ever known and continue to go to the sweet private school he'd attended since he was in diapers. But the day eventually came when I didn't have money for the mortgage or tuition. I wrote in my journal: "Dear God, I don't know how you are going to handle this, but I know you are. I need two thousand dollars. And I need it now. Thank you and amen." An hour later the phone rang. It was my sole remaining client. She said the strangest thing, "I don't know why, but I just feel you should send us an invoice for two thousand dollars." Of course, I did know why, but I just said, "I'd be delighted to do that." The next morning I wrote "THANK YOU" in huge letters and had a long chat with the Voice about gratefulness.

As my son's second-grade year came to a close, it was time to face the reality that private school was no longer an option. So I went on a hunt and found gold: a tiny magnet public school with just one third-grade class of twenty-five, high-performing students. There were only three openings, and they would be filled by a countywide lottery. All the principal could suggest was that I get my son's application in on time so his name would be in the pool.

I went straight to my journal. "Dear God, I found a beautiful, calm, peaceful school for my son. You know how much he has suffered. You know how frightened he is to go to a new school. Please. I trust you to place this precious child in a school where he will learn and be happy. I leave this in your hands." The morning after the lottery, the school called. My son was number one on the list of over three hundred children.

Eventually my savings were gone, and I had to put the house on the market. The last month in the house I faced a stack of bills with only $343 in my checking account. "Dear God," I wrote, "I know you hear me. I have no idea how you are going to do it this time, but I know—I know—you provide for us now and always. And all is well. Thank you in advance for the miracles you provide."

I blessed the envelopes and began. The lowest power bill I ever had was sixty bucks, but this one said I owed only $14.06. Despite my best efforts at conservation, the water bill always ran over a hundred dollars every two months. But when I opened this month's, I blinked at the amount due: $24.15. Gas was usually fifty-five dollars or so, but the amount due on this Mobil Oil bill was $13.13. The phone bill was only $22.98, while the cable bill was for the normal amount: $33.36. The garbage invoice was always for exactly $58.45, but this time, the amount due was zero; the statement said I'd paid double last month, but I didn't remember doing that. Finally, Visa. I knew this one wasn't going to be pretty. I took a deep breath, looked to heaven, and opened the envelope. Amount due: $ -39.09. In big letters it said: "Credit balance. Do not pay."

In the end I had $48 left in my checking account, enough to buy a week's groceries if I shopped carefully at the Greek produce stand. This time I didn't just write my thank-yous, I danced them. Up and down the hallway, laughing and twirling and singing my thank-yous with joyous yelps. "Thank you, God, thank you, God, thank you, God!"

Were these miracles? Coincidences? Delusions? If there were any doubt in my mind, it would soon be erased.

I wrote down my soul every day for three years. At first, I just complained about my problems and begged the Voice to fix them. But as I became more and more conscious of the direction and guidance I was receiving, I began to pursue deeper, richer questions—questions that probed my soul and lanced my deepest wounds. Profound answers, I discovered, came through profound questions: How did I create this mess? What was I thinking? How do I stop fighting with someone who won't stop fighting? What is the taproot of all my fears? Who are the negative voices inside my head? How can I banish them? What is my purpose? How can I build a conscious, joyful life? What is love? What is love really? What are my true vows, the vows that can never be broken?

That question about vows gave me profound pause. I wrote pages on end about vows—vows we make and vows we break. So many promises in life get broken. Are there really vows, I wondered, that I could never, *ever* break? I explored this question with the Voice for weeks. In deep dialogue, I concluded that vows aren't weighty promises made to fend off some undesirable future; no, true vows are words that articulate who I am, who I was, who I always will be. And if they are that—a description of who I am at my core—well then, I can never break them, can I? To break them, I'd have to stop being me. That left one big, big question to explore: who am I when I am fully me?

I asked. I made lists and pared them down. Is this true for me? Always? I played with the words. Can I say it better? More clearly? More succinctly? More powerfully? What words make my heart sing in recognition? Slowly the list narrowed to seven short declarations. When the seven felt sufficiently cooked, I typed them on a piece of paper and taped it to the wall. These, I told the Voice, are my vows. This is who I am, the real me, the whole me, the authentic me—the me I uncovered talking with you.

Janet's Covenant
7. Pray always
6. Seek Truth
5. Surrender, there is no path but God's
4. Come from Love
3. Honor myself
2. Live in Partnership
1. Unite to create Good

I looked at my covenant. It felt good to have my seven vows on the wall, reminding me daily who I am. But it didn't feel quite complete. I wrote about that: "Dear God, what's missing?"

Well, what was missing, I quickly learned, was the ceremony. When people declare their vows, they go through a ceremony—a wedding, novitiate, ordination—some kind of public declaration of their new commitment. That's what I needed—a celebration. I called a circle of eleven wonderful women to witness my covenant with Spirit on November 11, 2000. I read my vows, then we prayed and danced and sang and drank champagne and feasted on caviar and poached salmon. To honor the occasion and cement it for all time, I traded all the jewelry my ex-husband had given me for one gorgeous, dark orange Mexican opal ring with eleven tiny diamonds on each side and the word *seven* engraved on the inside.

My life had truly changed. The woman who had cried all day now moved through the world with vigor and purpose. The woman who had cowered in fear now confidently stood her ground. The woman who had been forced to sell her house now owned a beautiful townhome. The woman whose consulting practice had disappeared now wrote deep soul writing guides to heal broken hearts. "What happened?" people asked. Writing happened. Connection with the Voice happened. Deep questions and even deeper thinking happened. Willingness to change happened. Prayer happened.

Whatever it was, people wanted it—and I was on fire to share. A large Methodist church in Tampa invited me to address a divorce-recovery group on the topic of forgiveness. I had to think about what journaling exercises I had on forgiveness. I found one profound writing exercise of self-forgiveness and one that mentioned forgiveness in passing. But that was it.

At the group meeting, I shared my story. I showed the members pages where the Voice showed up. I read from some of my favorite sacred texts. I encouraged their active participation in writing deeply from the soul. I answered lots and lots of questions. When the time was up, they didn't want to leave.

Three months later, the church invited me to come back. The topic this time? Forgiveness. I went right to my journal: "Dear God, I don't have enough material on forgiveness and *you know it!* OK, I get it: the teacher needs to teach what the teacher needs to learn. Well, I'm ready. I just don't know how to do it. You show me how to forgive, and I will forgive."

Ask and ye shall receive—and boy, did I receive. Songs on the radio, articles in magazines, conversations with friends, even my book group's selection that month—they were all about forgiveness. I was swimming in a sea of forgiveness. I knew this bounty wasn't just for a good lesson plan. It was something more, something I needed to do, something missing in my life and my books. With that thought simmering in my head, I went to church the next Sunday. The minister opened her lesson with a Bible passage I'd never heard before: "It is someone who is forgiven little who shows little love" (Luke 7:47).

The moment I heard this verse, my heart knew what to do. I had to forgive my ex-husband—finally, totally, and completely forgive him. So he could love again. So I could love again. At long last, I wanted him to be free to love and be loved. As the minister spoke, I wrote the most beautiful and powerful prayer I'd ever written. When the service ended, I didn't move. I felt strange, woozy almost—like I was breathing different air. Something had definitely happened.

At five that afternoon, I pulled into a McDonald's parking lot to pick up my son from a visitation with his father. The moment my ex-husband saw me, he popped out of his car and started toward me. My stomach tightened, but there was no time to reach my cell phone. He knocked on the window. I lowered it four inches. His fist came flying in. I flinched back against the seat. Something fluttered to my lap.

"What's this?" I stammered.

"Half the dentist," he muttered.

I looked down. There was a check for thirty-eight dollars, exactly half our son's last dental appointment. According to our divorce agreement, my ex-husband was required to pay half our son's medical expenses. Until that day, he hadn't paid a dime and owed me thousands.

"Thank you," I called out to his receding back as he walked away.

The next morning I pondered in my journal about that check. "Dear God, why did he do that?" I turned back the page and looked at my notes from Sunday. There was my prayer of final and complete forgiveness—written an hour or two before he wrote that check. I never told my ex-husband I'd written that prayer, but from then on our relationship was less strained.

Our son, however, continued to struggle with visitation. Finally, in the spring of 2002, he looked his father in the eye, and said, "Dad, I'm not coming to your house anymore, and I'm not getting in a car with you again."

When my son told me what he'd said, I had two conflicting reactions. First, I was proud of him for finding the strength to speak his truth. On the other hand, if our son never saw his father again, how would they ever heal? And if they never healed, wouldn't that leave a gaping hole in both their hearts?

I talked the situation over with the Voice in my journal and realized what to do. I shared the idea with my son. He wasn't too crazy about it, but he let me call his father.

"Why don't you come here for visitation?" I said. "Come on Thursday and Sunday night for dinner."

My ex-husband came the very next Thursday. That first night we sat at the dinner table, looking at one another. This is strange, I thought, but somehow OK. The moment dinner ended our son scooted upstairs, ostensibly to do homework. The son and father probably didn't say five sentences to each other that night, but it was a beginning.

The next week, I told my mother my ex-husband was coming for dinner. She was appalled. "How can you let a man who tried to kill you back into your house?"

"Well, he isn't trying to kill me, Mom, and I'm not afraid of him, and I want our son to see that I'm not afraid, so he can stop being afraid."

She sighed, "I hope you know what you're doing, dear."

I told my sister. She hung up on me. I told my friends. One yelled at me. Most just shook their heads. My dearest friend tried to understand. She asked me why I was doing this. I told her I was doing it in the hopes that the pain could heal between our son and his father. "OK, good intention," she said, but "*how* can you do it?"

"Oh, that's easy," I said. "I have completely and totally forgiven him."

For fifteen months my ex-husband joined us for dinner twice a week, unless he was too sick to get out of bed. Then, on October 6, 2003, he died of a massive brain-stem stroke. October 6 wasn't just any day that year; it was Yom Kippur, the Day of Atonement, the highest holy day in the Jewish calendar—a detail I could not miss.

I called his best friend, and together we went to my ex-husband's business. We called his lawyer and banker and learned that the business was in dire financial straits and he would probably have gone bankrupt if he hadn't died. We met with his few remaining employees and arranged their last paychecks. When the office was finally empty, I sat in the dust and started to go through his files. Faced with six rusty, five-drawer file cabinets, packed to the gills with unorganized and often unlabeled files, I began at the bottom, pulling out each file, reading it, and trying to figure out what to do with the papers inside.

Stuffed in the back of the third drawer, I found a thick file labeled "Life Insurance." Our divorce agreement had required him to carry $250,000 in life insurance for our son, but the papers showed he had let his life insurance lapse.

Then, through my tears, I also saw that the week I had invited him to start coming to our home to see his son, he had begun a contentious battle with his life insurance company. Although it had taken him six months and $7,800 he did not have, not only did he get his insurance reinstated, but he had also increased it and named me beneficiary. When I received that check for $322,000, I knew I was holding tangible proof of the power of forgiveness. And, just in case I missed the connection, the check was dated November 11, 2003—three years to the day after my covenant celebration.

I wonder sometimes what my life would be like if I had not engaged the Voice in deep soul dialogue. Would I have safely navigated the terrors of my divorce? Would my heart have healed? Would the same miracles have happened? Would I have completely and totally forgiven my ex-husband? Would he have increased his life insurance and left it to me? I can't rewind the tape of my life and then play it forward in a new scenario without deep soul writing, so I guess I can never really know. But I'm fairly certain the answers would all be no.

Writing from deep within my soul is now ingrained into my daily spiritual practice. It is how I meditate and how I pray. It is how I solve problems and how I learn. It's where I mourn and where I express joy and gratitude. It is who I am.

And it may well be who you are, too. After all, it is no accident that this book has come to you. In the big scheme of things there are no accidents, only divine appointments. My divorce was the worst experience of my life. It was also a divine appointment—an appointment with destiny, with the Voice, and with my self. Without the divorce, I might never have discovered the Voice, and without the wisdom of that Voice, I simply could not have the life I have, the work I have, and the joy I feel today.

We humans are an odd bunch. We are not very likely to turn to the divine in times of love and plenty, but let those winds of destruction come, and we can't fall to our knees fast enough. If you are the kind of soul that needs a setback to force you to turn inward, well, the universe, I'm sure, will be happy to comply. But here's a little fact that might warm your heart: you don't *have* to experience a trauma to receive that invitation. It's a standing invitation, open to all. Accept it, whether you are currently in good times or bad, and you will experience direct and immediate access to divine consciousness. Accept it, and you will hear and see the Voice. Accept it, and you will receive the wisdom and miracles your heart is longing for.

How do you accept? That's easy. Set your intention to connect with the extraordinary Voice within, pick up a pen, and begin.

What Is Writing Down Your Soul?

Is This Journaling?

JOURNALING, THE WORLD SEEMS TO AGREE, is a good thing. Pick a book—any book—in the self-improvement section of the bookstore, and you'll be hard pressed to find one that doesn't recommend journaling. Christina Baldwin jumpstarted the trend in 1990 with *Life's Companion: Journal Writing as a Spiritual Quest*. Next came Julia Cameron's 1992 classic, *The Artist's Way*, with directions to write three "morning pages" a day to heighten creativity. A few years later, Sarah Ban Breathnach followed her bestselling *Simple Abundance* with the *Simple Abundance Journal of Gratitude*. Since then, it seems every blockbuster from Rick Warren's *Purpose-Driven Life* to Stephen Covey's *Seven Habits of Highly Effective People* to Joel Osteen's *Your Best Life Now* arrives with a matching journal. Some books even come out as stand-alone journals. Phil McGraw (Dr. Phil) wrote *Life Strategies Self-Discovery Journal* to give readers the opportunity to experience insights rather than just read about them.

And don't think journaling is limited to the "softer" arenas of spirituality and personal growth. Search for "journal" at any online bookseller, and you'll find journals for everything from dating and diseases to weight control and wine tasting. Even the data-driven world of finance and business sees the value of journaling. In *The Success Principles*, Jack Canfield, coauthor of the mega-bestselling *Chicken Soup* series and the nation's foremost success coach, recommends journal writing:

> Many people have their greatest success accessing intuitive information through journal writing. Take any question that you need an answer to and just start writing about it. Write down the answers to your question(s) as quickly as they come to you. You will be amazed at the clarity that can emerge from this process.

Other professionals encourage journaling, too. Therapists often recommend journaling to deepen the insights and strengthen the gains made in therapy. Hypnotists encourage their clients to journal. And spiritual directors from many traditions counsel seekers to explore their spiritual life on paper. From teenagers spilling their hearts into their diaries to weight watchers recording every bite, millions of people journal in one form or another.

But not all forms of journaling have equal impact. There are several things that distinguish writing down your soul from typical journaling.

The first is **intention**. While hundreds of books and thousands of teachers extol the benefits of journaling, they rarely mention the critical first step—the thing you do *before* you pick up a pen: set your intention. It is the energy of intention that puts everything in motion. When you begin with a clear intention to access the Voice of wisdom within, you let the universe know that you are ready to open two doors in your soul: the door into your deepest self and the gate to the cosmic divine. That's a powerful combination and an unmistakable message. And the universe responds—always. So by setting your intention to open your soul to divine dialogue you elevate the act of writing to a place regular journaling rarely goes.

The second difference is **purpose**. This kind of writing has a singular, deeply personal focus. It is about you, your life, your concerns, your fears, your aspirations. It's about discovering and giving voice to the secrets buried deep in your soul. It's about asking questions until you uncover the ones you've never asked. Its purpose is to deliver the guidance you need *right now* to live the full, rich life you are here to live. Write with that purpose in mind, and you will uncover a trail of answers—not just any answers, but *your* answers.

Writing down your soul, as you are about to discover, also has a unique **process**. In the next pages, you will learn to write in a way books on journaling do not discuss.

Another big distinction between writing down your soul and journaling is **commitment**. There's a rub in this kind of writing, a rub that takes it well beyond the self-exploration of standard journaling. The rub is: if you ask the universe for guidance and receive it, you are a bit beholden to do something with it. You can't hear the Voice of wisdom and then say, "Hey, thanks for the advice, but

I think I'll just keep doing things my way." Why open that door in your soul and then pretend it isn't open? Why walk through that magical gate and then pretend you didn't? In deep soul writing you search the reaches of your heart, tell your story, ask your questions, hear your answers, and receive your guidance. What you do next in response to that guidance is up to you, but you can't really ignore it.

Set your intention, write with a purpose, follow this process, and make a commitment to use the wisdom received. Do all these, and you are no longer journaling—at least not in the usual sense. You are writing in and with and through your soul. You are connecting with the Voice, asking for guidance, and receiving it. You are opening yourself to the grace and gifts of the universe. You are changing your life.

Is It Meditation?

When we hear the word "meditation" we typically visualize someone sitting on the floor, legs crossed, back straight, breathing in and out, silently seeking an empty mind. For years I tried to meditate this way. I tried guided meditation, silent meditation, and chakra meditation. I practiced with mantras, mudras, sounds, chimes, colors, and, of course, breath. But I could never get my mind to do that clearing thing. Riggedy-raggedy thoughts were always there prodding and poking me. "Acknowledge them," the instructor would say in a warm, liquid voice, "and let them go by." Trust me, I'd think, these thoughts aren't going anywhere. They're too busy reminding me of all the things I have to do and all the problems I have to fix.

I take no credit for discovering deep soul writing. I stumbled upon it. Out of sheer desperation, I picked up a pen and wrote,

"Dear God," at the top of the page. Immediately all my fear thoughts lined up to be heard. They jumped through my pen and onto the page where I could see them and where the Voice could show me how to heal them. For three years, I told my story and asked for guidance. And, for three years, guidance came—every day.

The *Oxford English Dictionary* says meditation is "the practice of profound spiritual or religious reflection or mental contemplation." Writing down your soul is certainly profound. It is definitely a spiritual practice, and it is probably the deepest reflection you've ever experienced. And your mind is totally engaged, but so are your heart and your soul and your body. So this kind of writing is a kind of meditation—and more. This kind of written meditation meets you where you are right now, no matter what's on your mind. It is time with your best friend and the wisest counselor in the universe, rolled into one. This kind of meditation is an intimate, personal conversation that can't be explained or even really shared. But you know it's real, because there it is on the page—your own personal conversation with the divine Voice within.

Is It Prayer?

I don't see how the answer to "Is it prayer?" can be anything but yes, because prayer is conscious connection with the divine. When you write this way you are certainly conscious, although you are also *not* conscious. By that I mean you're not limited to the conscious level. As you write, you dive below the conscious to thoughts and feelings you didn't know you had, and you soar above the conscious to experience real understanding, safety, and peace. This kind of writing is definitely a connection. You quickly recognize when you are in touch with something within yourself that is beyond yourself. Call it by any of the names we assign to the divine or call it the Voice. Call it whatever you want. The name isn't what matters. The point is that you know you are connected to something *more*. At some point even the word *connection* becomes insufficient to describe the experience. *Communion*, perhaps, would be more accurate.

So is writing down your soul journaling? Yes, in the sense that you are writing in a journal, but deep soul writing is so much *more*. Is it meditation? Yes, but it is also something *different*. Is it prayer? Yes, but it is a *new kind*. It is all these things and more. It is what happens when journaling becomes meditation becomes prayer.

When you engage in this kind of writing, you enter into a continuous loop of communication between you and the Voice within you. You write, and the Voice listens; the Voice writes, and you listen. It's that simple—and that mysterious. The Möbius strip is the ideal symbol for this kind of exchange with the Voice because it has no end and no beginning, no inside or outside, no stop or start.

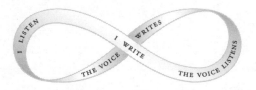

Perhaps we should not try so hard to answer the question, "What is writing down your soul?" In the end, each person will find out for him- or herself. Your answer will not be the same as mine. Your experience will not be the same as mine. And maybe that's just the way it should be.

What Do I Need?

WRITING DOWN YOUR SOUL can be an occasional relief valve or an ongoing conversation. If you just want a taste of the Voice, all you need is paper, pen, and the impetus to sit down and write. But if you want to open that door and keep it propped open for constant and immediate access to the extraordinary Voice within, do a bit more. Here's everything you need to do to set up a spiritual practice of writing that gets richer and deeper over time:

- Create a schedule
- Stop
- Get a journal, notebook, or other paper
- Pick a pen
- Make yourself available

- Create a sacred writing space
- Secure your journals

It's easy; you could do it all in a couple minutes. It's also amazingly inexpensive. Your total investment could be as little as a few dollars. Your total return, however, could be priceless.

Create a Schedule

Just like the wise financial practice of "pay yourself first" creates huge financial benefits over time, the wise spiritual practice of "tell the Voice first" generates huge spiritual benefits over time. Pick a slot in your day to talk with the Voice. Block out ten to fifteen minutes—more when you are under stress. There is no right time or wrong time to write. The time of day doesn't matter; making a commitment and keeping it does.

Most deep soul writers report that they write in the morning before their day begins, even if that means getting up a few minutes early. It's the only way, they say, that they can be sure they'll have uninterrupted time to write.

I love this idea, but I'm too muddle-headed and uncoordinated first thing in the morning. Instead, I like to have my coffee, read the paper, and attack the Sudoku puzzle before I go upstairs to my office. There, I repeat my personal covenant and say my daily writing blessing out loud, blessing my work, my life, and my hands. Then I sit in my writing chair and have at least a ten-minute written conversation with the Voice. When I'm finished, I move to the computer for my professional writing. I'm always tweaking my process, but that's basically the schedule that works for me.

A few writers report that they are able to write during a scheduled hole in their day, such as lunch or mid-morning break. This plan works well when the break is consistent and you can count on privacy. It doesn't work if you're prone to work through your break or if you are concerned that someone will read over your shoulder or ask you what you're doing. If your home life is too chaotic, you might want to look at your work schedule and see if you can create a regular writing break.

Many people in Writing Down Your Soul workshops report that they write just before they fall asleep. I've always known that this kind of writing has a calming effect, but Nancy, a professional woman in her mid-thirties, surprised me and everyone in our deep soul writing class when she blurted out, "I can't wait to journal every night! It completely settles my chatterbox mind. When I write, I'm done, and I don't have to think about my problems any more. For the first time in my adult life, I'm sleeping through the night! I *love* this practice." I thought I understood the power of writing, but this writing-as-a-sleeping-pill idea was a new and amazing discovery.

Donna Vernon breaks her writing schedule in two. When Donna first came to a Writing Down Your Soul workshop, she was a massage and colon therapist, a career that "wasn't working." Today, after rewriting her life, she is the founder of a Web-media firm disseminating information on health and complementary medicine to the world. "I continue to write every day," Donna told me ten months after the class. "I say my verbal affirmations in the morning, describing my life as I want it to be, then I write whatever comes. At the beginning, it took me twenty minutes or more because my life was so out of sync with what I wanted. Now it's pretty short, more of a to-do list. And I do a gratitude journal at night. I write what

I accomplished, how it fits in line with the new self I'm creating. I mention all the gratitude I have for the people I talked to, the situations I was in, even the challenges I faced that day." Donna has certainly found her perfect writing schedule.

The key to a successful schedule is to write at approximately the same time every day. Don't say, "I'll write when I can." Trust me, you'll never find the time. It isn't that you don't *want* to or even *try* to. The problem isn't you; it's the worried little critic who lives deep inside of you who is scared to death of what's going to happen when you and the Voice start having real conversations. When that little critic gets wind of change, he or she can be counted on to start tossing up all kinds of scheduling blocks. If you find yourself saying, "I'll write *later*," or "I'll write *after I take care of* _____" (fill in the blank with everyone who needs you to stop what you're doing and do what *they* need), your inner critic is doing a good job deflecting you away from something that could change your life. If you find yourself skipping your writing time more than you make it, it doesn't mean you're lazy or weak or too busy; it means your inner critic—not you—is in charge. Get back on top by making a schedule and sticking to it.

Cost: $0

Stop

Writing is not something you *have* to do; it's a gift you give yourself—the gift of stopping. Stop for a just a moment and step off that busy, go-go, do-do, get-get train that we all ride all day, all week, all year. Stop for just a few minutes and talk with the Voice. Don't worry that the train will speed on ahead without you. Not only will you have no trouble getting back on, but you'll also re-board with new answers, new clarity, new energy, and a smattering of fresh, new hope.

Cost: $0

Get a Journal

Any writing material will do. At first your journal may be a stray legal pad floating around the house or a child's leftover composition book. If the urge to write comes while you're stuck in the car, your journal may be something as ridiculous as the back of a fast food napkin. (I've learned the hard way that it's wise to keep a small spiral notebook in the glove compartment.) The medium really doesn't matter, but eventually, you'll want a nice blank book. Retail and online bookstores are full of journals—so many, in fact, that the options can seem overwhelming.

When I bought my first journal, I had one criterion in mind—cost. The cheapest blank book at my local bookstore was a no-frills, 11" x 8 1/2" notebook with a plain, black cardboard cover and a hundred and fifty completely blank pages. I highly recommend this kind of ultraplain, low-cost, big-size blank book. It gave me lots of room to yell and vent and blow off steam in big ugly letters. And it was so cheap that when I plowed through the first journal in two weeks, I didn't care. I just set out to get another. After I filled a dozen of those, I was ready for something smaller and slightly more beautiful.

I loved going to the bookstore to choose a journal. I experimented with the binding and size. I discovered the joy of writing in a spiral notebook that opens completely flat and the nuisance of trying to write in a journal that's too small. I bought skinny journals, fat journals, journals with ribbons, journals with sayings, journals with white paper, beige paper, lined paper, blank paper—journals of every dimension, description, and price. Once, when I was feeling really indulgent, I bought a sumptuous blue suede journal. I didn't want it to ever end. I don't buy journals anymore; my son has discovered that they are my favorite gifts from him. His choice is a little window into his view of me at that moment, and his inscription means as much to me as the conversation the book will contain.

When you go to the store to buy your journal, consider the possibility that you may not be selecting the journal; the journal may be selecting you. Theresa had this experience. She held up her journal and shared this story with her Writing Down Your Soul group:

> After class I went to two stores and just couldn't find the right journal. So I stopped at the corner drugstore and bought a simple spiral notebook and started writing in it, thinking I'd get a real journal later.
>
> When I was a kid, I always drew daisy flowers for my grandfather with big loopy petals in bright colors. We were very close, and I miss him greatly. A few days after I started writing I had a dream. In the dream my grandfather hands me a piece of paper and says, "This is very very important for you to know." I look at the paper, and it's a flower drawing just like I used to make for him. The next morning when I picked up my journal to write, I looked at the cover and on it was a big loopy colorful daisy—just like the picture in my dream. When I bought this notebook, I had no memory of those flower drawings. I did not consciously choose it for that reason. But now when I write in

it, I feel connected to my grandfather. I'm so glad I bought this notebook.

Have fun selecting your journal, but don't go overboard. If the journal costs too much, you won't give yourself permission to write as long or as fully as you need to—or as long and fully as the Voice wants to. I once purchased a heart-stopping hand-bound journal in a California art gallery with a collage on the cover and handmade paper inside. It was beautiful to look at and sensuous to hold, but I could never bring myself to write more than two pages at a time. The point, I had to remember, is the conversation, not the container.

Cost: $0 for paper found around the house

$2–4 for an inexpensive spiral notebook

$10–20 for a really nice journal

Pick a Pen

Your writing instrument can run the gamut from the half-chewed pencil next to your phone to the cheap ballpoint your bank gave out last winter to a luxurious new fountain pen. The way I see it, my pen is the physical vehicle for the Voice of Spirit, so I treat it with reverence.

After writing for years, I fell in love with a lightweight Mont Blanc "Generation" fountain pen. It cost over two hundred dollars, which was an extraordinary indulgence at the time, but once I held it in my hand, I was a goner. I carried that pen with me wherever I went: church, concerts, speeches—anywhere I might hear something I'd want to remember or talk over with the Voice.

Then, four years after I'd bought it, the pen disappeared. I was heartbroken. I tried to write with something else, but it just wasn't the same. When I went to the Mont Blanc store in Tampa to get another, I was told the model I had was no longer produced. I looked at every pen in the store, but none of them felt right. Taking pity on me, the salesman submitted a worldwide search for a Generation model just like the one I'd had. It took eight months, but at last he found one for me. When he gave it to me, I wrote "Dear God" over and over again in big, elated script across a whole page of the store's sample writing paper. I still take the journal with me everywhere I go, but my beloved fountain pen stays home.

    Cost: $0 for whatever is floating around the house

        $3–5 for a nice ballpoint or rollerball

        $? for a pen that grabs you by the heart and won't let

        you go

Make Yourself Available

Once you make it clear to the universe that you are open, information and ideas will come through, whether you are sitting in your writing chair or not. Your dreams will become laden with symbols and stories you'll want to discuss and interpret with the Voice. And you will probably be awakened in the wee hours by words that simply demand to be written. I can gauge how intense or important something is by the wackiness of my dreams and the not-so-gentle nudges to "wake up and write." Keep your journal or another notebook beside your bed, and carry a small, pocket-size notebook everywhere you go.

    Cost: $0

Create a Sacred Writing Space

Just as you can pray anywhere, you can write anywhere—in bed, in the car line, in a coffee shop, at the kitchen table, in your back yard, at your desk. The actual location is irrelevant. Yet we gravitate to sacred places such as churches, synagogues, mosques, shrines, and meeting rooms because it feels good to pray in a setting we deem holy. And so it is with writing. You can write wherever you are, but you can also create a sacred writing space that reflects you and your relationship with the Voice. This place can be as small as a designated drawer in your nightstand or as large as a whole room. All you really need are quiet, privacy, light, and space for your journal and pen.

If you want to create a deeply personal writing space, begin with a comfortable armchair. Put a table next to it with your journal and pen, a good lamp, and the sacred texts or spiritual books you enjoy. Next, add whatever enhances your experience: artwork, talismans, candles, holy oils, prayer beads, rocks, crystals, feathers, photographs, prayer rugs—anything and everything that pulls your consciousness inward and upward. Got an ottoman? Toss that in, too. You want to be comfortable when you write, so your whole being can be completely focused on the writing.

But heed this warning: do not delay writing while you search for the "perfect" items for your "perfect" writing space. This practice is not about the place; it's about communion with the Voice. And the Voice doesn't really care about the room where you are sitting; the Voice just wants to talk. If you find yourself spending more time working *on* the space than you do writing *in* the space, recognize that it's not you consciously avoiding writing, but once again, that little inner critic making sure you don't stir things up. Push him or her aside, sit down, and start writing wherever you are.

Cost: $0

Secure Your Journals

Your journals are sacred. They are a record of your most private conversations with the Voice. They are for you and the Voice, and *only* you and the Voice. Just as you would not give another person a word-by-word transcription of your prayers, you do not want anyone to read a transcript of your written conversations with the Voice. So plan a place to keep your soul's notebooks private.

A couple months after I started divorce proceedings, my husband broke into our home. He took only one thing: my journal. Luckily I was writing with such intensity and speed that only the Voice could read the furious scribbles on the page. Eventually my husband gave the journal to his lawyer, who gave it to my lawyer, who gave it back to me. As soon as I got it back, I sat down and wrote in it. But I did not leave it on the table next to my writing chair in the living room anymore. I hid it among the files in my office.

At Writing Down Your Soul workshops the group invariably talks about how to keep journals from prying eyes. A few, typically those going through relationship trauma, realize they have to write outside the home or write only when they are alone. And then they carefully hide their journal, or, if necessary, they write on loose-leaf paper and put the pages through a shredder.

Not sure what you need to do? Just ask yourself if your partner or children or anyone else in the house would look in your journal. If the answer is yes or if you're not sure, stay on the side of caution and protect your journals. Do not, however, use privacy as an excuse not to write. People going through relationship transition need to write as much or more than anyone else.

Older writers sometimes talk about not wanting relatives to find their journals after their death. At a recent workshop the group proposed an interesting solution to this dilemma: select a date and hold a bonfire "party" with a few special friends (like the people in your Writing Down Your Soul group). At the party, carefully and slowly burn the pages in a copper bowl, fireplace, or grill. (Be sure you know what you're doing and can stay in control of the fire before you go down this route. A shredder may not be as dramatic, but it's just as effective.) An alternative for people who trust their families: leave a request in your will asking that your journals be destroyed without being read. I know I could put that in my will, and my son would honor the request.

The reason this topic of how to destroy journals comes up is that once you start this dialogue, you can't stop. Let's say you write an average of three pages a day—that's twenty-one pages a week. If your journal holds 200 pages, it will last for ten weeks, or two to three months. At that rate you'll fill at least four journals a year. If you write for twenty years— well, you get the picture! That's a lot of journals piled in the closet or in the basement or somewhere. (Confession: I have so many—and I'm not ready to let any of them go—I had bookshelves built into my office closet.) So if you are uncertain of how private your journals are, stop for a moment and decide how to protect them. Otherwise, write on.

Cost: $0 if you live alone or with people who respect
your privacy
$0 if you burn the pages (maybe extra for some coals
and lighter fluid)
$25 or so if you get a shredder

Who or What Is Listening?

The whole time I was embroiled in intense daily conversation with the Voice, I never gave any thought to whom or what was at the other end of the conversation. I was too focused on me and what I needed to learn or know or do to survive. I was in an intimate relationship, having rich dialogue with "Dear God," whoever or whatever that is. But when I started teaching other people to write, I had to consider this question, because addressing the Voice directly is a key element in writing down your soul. At the time, I called my workshops "Dear God," but I told my fellow writers to address the Voice however they wanted. The power isn't in the name, I told them; it's in your intention. Use the name that speaks to you.

And that worked pretty well until I submitted a proposal for this book under the title *Dear God.* Jan Johnson, publisher of Conari

Press, said, "There are a lot of people for whom the term *God* is not automatically warm and welcoming. Why don't you go off and think of something more inclusive for the title."

After a few days of wrestling with this problem by myself, I gave up and sent an email to everyone who attended one of my workshops. I asked a short, simple question, "How do you address the Voice?" That question generated a wellspring of beautiful names and propelled me into a search through ancient spiritual traditions and modern science.

This problem of what to call the divine is not unique to our time. From the beginning, humanity has struggled with what to name the unnamable. *God* may be the going term in the West—tossed around pretty casually in jokes, advertisements, prayers, and curses—but it's just an old Germanic/Dutch/Saxon word for "supreme being." Before there was the European word *God*, there was the word *Allah*. And before *Allah*, there was the word *Yahweh*. And before some societies got behind the idea of one god, there were hundreds of names for hundreds of gods—Egyptian gods and Greek gods and Roman gods, Celtic gods and Norse gods and Russian gods, Hindu gods and Zulu gods and Chinese gods, Aztec gods and Native American gods and aboriginal gods. Wherever humans dwelled, there were gods—and those gods all had names.

A few soul writers saw my inquiry as an attempt to be politically correct and sent rather terse responses that *God* is good enough for them, but most of my writing friends confirmed that Jan Johnson was right. There is indeed a growing sentiment that the Germanic term *God* is too limiting, too patriarchal, too judgmental, too *something,* and somehow pushes people away—which, of course, defeats the very purpose of calling on the name of the divine.

Who or What Is Listening? 47

Many people today are actively searching for gentler, more inclusive, more compelling names for divine consciousness. In the DVD phenomenon, *The Secret*, the speakers use the term *God* a few times, but they also use *Infinite, Power, Law, One, Creator, Holy, Genie, All, Energy, Spirit, Spiritual Being, Source, Source Energy, God Force*, and *Infinite Possibility*. The word they use most often is *Universe*, which makes sense because the divine (however it is called) created the whole shebang. As the often-quoted Reverend Michael Beckwith of *The Secret* likes to say, "God is not in us; we are in God." In other words, all of the universe is divine consciousness. But, as several of my writing buddies pointed out, it can be tough to have an intimate conversation with something as big and cold and distant as "the Universe." Although none of them felt they had found the perfect name, they humbly offered these suggestions:

Spirit, Higher Spirit, Great Spirit, Holy Spirit
One, Holy One, Divine One
Creator, Source, Lord, Almighty
Presence, Essence, Life Force, Light
Great Mind, Divine Mind, Master Mind
Being, All Being, Energy, All That Is, Everything
Heart, Heart of Hearts
Mystery, Great Mystery, Great Wonder
Beloved, Joy, Truth
Father, Mother, Father-Mother God
Friend, Witness, Listener, Guide

If my little survey had been a democratic vote, *Spirit* would have won. More writers mentioned *Spirit*, or a version of it, than any other name.

Since receiving my fellow writers' emails I have two new favorites: *One* and *Friend*. Imagine having an internal conversation with the One—the alpha, the omega, the all. I like that. "Dear One." *One* seems like a perfect name for the divine because it captures the ancient concept of the oneness of god and at the same time communicates the very modern idea, articulated so beautifully by Beckwith, that we and the divine are one.

And then, there's *Friend*. How sweet is that? "Dear Friend." The divine is certainly that—the Friend who listens fully and will not let you get away with anything. The Friend who gently and lovingly points out the errors in your thinking, the discrepancies in your behavior, and all the ways you get in your own way and block your own good. Yes, the divine is definitely our complete and perfect Friend.

But before I settled on a new name, I decided to do some exploring. I started with the sacred text of my childhood—the Bible—but I didn't stop there. One name led to another and another and another. I had a delightful time exploring humanity's apparently endless names for the divine. In the end, of course, you and the Voice are the only ones who can choose the name for your time together, but perhaps my explorations will spark some divine ideas.

I started with the Tanakh, commonly referred to as the Old Testament. According to the first book of the Bible, when Abram was ninety-nine years old, "Yahweh appeared to him and said, 'I am El Shaddai'" (Genesis 17:1).

El translates as "God." *Shaddai* translates as "almighty" or "destroyer." I looked up *El Shaddai* and learned that there are not one, but seven holy names of God in the Jewish tradition: *El*, *Elohim*, *Adonai*, *YHVH* (which technically has no vowels but is still often pronounced *YAH-weh*), *Ehyeh-Asher-Ehyeh*, *Shaddai*, and *Tzevaot*. And, if that wasn't enough to choose from, there is another less well-known tradition discussed in *She Who Dwells Within* by Lynn Gottlieb. Rabbi Gottlieb opened my eyes to another dimension of the divine.

> The term *Shekinah* is an abstract noun of feminine gender derived from the Hebrew root Sh-Kh-N, meaning "to dwell" or "to abide." The word *Shekinah* first appears in the Mishnah and Talmud (ca. 200 CE), where it is used interchangeably with YHVH and Elohim as names of God.

A feminine name for the divine—that certainly opens a whole new stream of possibilities. Just think of all the great female goddesses, like Athena in Greek mythology, Brigit in the Celtic tradition, and Isis from Egypt. Start looking up goddesses, and you will be overwhelmed by the variety. It seems that there is a goddess for every inch of earth, every human need, and every divine mystery.

I wondered about Moses. Now there was a guy who talked to the divine a lot; surely he addressed it by a name. I started reading Exodus and laughed out loud when I got to the third chapter. The divine being has just told Moses, who was nothing more than a slave in the eyes of the Egyptians, to go to the Egyptian pharaoh, who was nothing less than a god, and tell him to let the Israelites go. Moses is a bit taken aback.

> Look, if I go to the Israelites and say to them, "The God of your ancestors has sent me to you," and they say to me, "What is his name?" what am I to tell them? God said to

Moses, "I am who am . . . This is my name for all time, and thus I am to be invoked for all generations to come." (Exodus 3:13–15)

Sounds pretty emphatic. There's just one problem: what does "I am who am" mean? Scholars have been debating that for a few thousand years. And writing, "Dear I Am Who Am" seems a bit awkward. So I kept reading. A bit later in Exodus, Moses, complaining about how horribly the pharaoh is treating the Israelites, whines, "When are you going to do something about this?" The divine answers, "I am Yahweh. To Abraham, Isaac and Jacob, I appeared as El Shaddai, but I did not make my name Yahweh known to them" (Exodus 6:2–3). Well, how do you like that! The divine changes its name every once in a while. Clearly, its name, whatever that name is, is not written in stone.

That got me thinking about the Shema, the statement of belief that orthodox Jews repeat daily: "*Sh'ma Yisrael Adonai Elohaynnu.*" You'll recognize it in English as "Hear, O Israel, the Lord your God is One" (Deuteronomy 6:4). How interesting that such an ancient name, *One*, is one of my two new favorites in the twenty-first century.

In the Muslim tradition, the divine has not one but ninety-nine names. The first is *Allah*; the ninety-ninth is *As-Sabur*, meaning "the patient" or "the timeless." A holy practice is to recite all ninety-nine names in order. Surely, that would cover all the bases.

Then there's Jesus calling upon "our father" (Matthew 6:9, Luke 11:2). At least "father" is what it says in all the modern English translations of the Bible. But the Bible wasn't written in English. The English version was translated from Latin, and the Latin version was translated from Greek. But—surprise—Jesus didn't speak Greek. He spoke Aramaic, a Semitic language, as did his listeners. As a child I remember hearing that when Jesus called upon God, he said *Abba*, which is closer in meaning to "daddy" or "papa." I always thought the abba/daddy translation sounded sweet and told me something about Jesus' intimate relationship with God. But according to Neil Douglas-Klotz, a scholar of the Peshitta Gospel—believed by many Eastern churches to be the earliest gospel and the only one written in Jesus' native Aramaic—when Jesus taught his followers to pray, he told them to say "*Abwoon d'bwashmaya.*"

Abwoon d'bwashmaya cannot be translated into one English word or phrase because Aramaic is a Middle Eastern language with three layers of meaning, all reverberating at the same time: literal, metaphorical, and mystical. If English conveyed such multiple meanings, a simple sentence like "I opened the door" might carry three simultaneous meanings: I opened a physical door in the wall (literal). I entered my subconscious (metaphorical). And the angels were all around me (mystical). Imagine what would happen if that triple-layered sentence were then translated into a one-layered language. Only the literal meaning would come through. The metaphorical and mystical meanings would be lost. Aramaic is not a dead language, so this is not linguistic speculation. People in Syria still speak Aramaic, and it continues to be the language of several eastern churches. Jesus' listeners undoubtedly understood the meaning of his words on several levels. And thanks to Douglas-Klotz, we can too. In his small, stunning book, *Prayers of the Cosmos*, he explains these multiple meanings and offers several rich translation possibilities for "Abwoon d'bwashmaya."

O Birther, Father-Mother of the cosmos!
O Thou! The breathing life of all
Respiration of all worlds
Source of Sound
Radiant One
Name of names
Wordless action, silent potency

Clearly, Jesus was saying a whole lot more than "father."

Travel further east and you come to the Sanskrit *om*, venerated not so much as a name but as a mystical syllable conveying the sound or essence of the divine. Hindu, Buddhist, Sikh, Jain, and Zoroastrian believers all chant some variation of the sacred sound of om.

In the Hindu tradition, there appear to be many gods. But they are all aspects of one supreme being that has three important tasks: to create, to preserve, and then to destroy and create again. If you want to use a Hindu name, write to Brahma if you want to create something, Vishnu if you want to preserve what you have, and Shiva if you want things to change.

I couldn't leave my research without considering the non-god names for cosmic consciousness. By that I mean the scientific names that attempt to capture the essence of the numinous without the trappings of religion. Names like *the Gap*. The first time you heard Deepak Chopra say that, weren't you befuddled? But at the same time, didn't it make sense? Somewhere in grade school, we learned we are made of atoms, and inside those atoms are protons and neutrons, and orbiting around those are infinitesimal electrons. And what is between all those minute particles? Space—lots and lots of empty space. It seems a bit bizarre, but we are not the dense flesh we see in the mirror; we are rapidly vibrating globs of space. In other words, we are walking, talking gaps. And our little gaps are connected to, or part of, or one with the Big Gap. Sounds a bit like a scientific explanation of "created in the image and likeness of God," don't you think?

For those who want to take the search for a non-god name for divine consciousness a step further, read—or try to read—something about quantum physics. Put very simply, quantum physics says everything is all one big, connected energy field called the quantum field. We are part of that field, obviously, so if we call the endless quantum field "God," that makes us little "godlets." So when we write to the divine, our small godlet expressions of energy connect with the big God expression of all energy, all creation, all life. I can understand this concept intellectually, but writing to "Dear Quantum Field" doesn't quite work for me.

After two weeks of exploration, I came to this conclusion: the more you search and the more you read, the less you know for certain about *the* divine name. And, in the end, that uncertainty is a good thing. The truth is, there is no one perfect name. Think of divine consciousness as a huge disco ball with thousands of mirror squares. Each small mirror is a different name. No one square or name is wrong; each is a tiny reflection of our human understanding of the nature of the divine. Taken together, the glowing ball radiates the full glorious light of the cosmic divine.

In the end, the only name that matters is the one *you* choose, the one that makes your heart sing. It's your conversation with "I am who am"; begin it as you wish. Use the name you've always loved or select a new one. Or experiment with different names. Use Jewish names, Islamic names, Aramaic translations, Hindu names, or the Sanskrit om. Use Christian names, Pagan names, or a name you make up altogether. That is what the great mystic Rumi did:

It Works

Would you come if someone called you
by the wrong name?

I wept, because for years, He did not enter my arms;
then one night I was told a
secret:

Perhaps the name you call God is
not really His, maybe it
is just an
alias.

I thought about this, and came up with a pet name
for my Beloved
I never mention
to others.

All I can say is—
it works.

If you're uncertain, ask the Voice to reveal its alias. Put this question in your soul journal: "Dear _____, What is your name, your alias? What shall I call you?" Then write anything and everything that pops into your mind. Do not judge or even think about anything that shows up on the page. Just keep writing for ten minutes. Don't feel any pressure to find the "right" name. Use whatever works for you and know that you and the Voice can change it any time you like.

Whatever the name, I am confident of one thing: the Voice answers. Always.

Dear _____,
What is your name, your alias?
What shall I call you?

Choosing a name is a delightful exercise, but for some people it still leaves a nagging question: Who is the Voice? Is the Voice some form of divine being or consciousness? Is it an angel, guardian, or guide? Is it some sort of divine spokesperson? Or is it a higher form of yourself talking back to you from a future, more advanced state? The answer is yes; the Voice is all that and more.

I don't mean to be glib. The Voice, you see, is a paradox. It is the largest consciousness, yet it squeezes itself through the tiniest tip of your pen. It's the massive noise of creation, yet you hear it as the faintest sound deep within your soul. It is cosmically profound, yet it speaks with you about very ordinary things.

Does that paradox frustrate you? Don't worry. The Voice will reveal itself to you over time. Your understanding will evolve and your trust and intimacy will grow. In the end, the Voice is who you believe it to be, where you perceive it to be, what you know it to be. Be at peace with that and enjoy the conversation.

Why Write?

WHY WRITE? THERE ARE, AFTER ALL, other ways to access internal wisdom. You can pray. You can meditate. You can work with a mental-health counselor or hypnotist. You can participate in religious rituals, visit a spiritual director, or see a shaman. You can talk with a loving friend over a glass of wine or a cup of coffee. And the easiest—and perhaps most accessible—way is to dream. There are many routes into inner consciousness, so why take the time to write?

There is only one reason: writing works amazingly well. If you want to engage in a vibrant conversation with the wisdom that dwells just a hair below your conscious awareness, write.

But why does writing work so well? Today, there are scientists in fields as diverse as psychology, physics, biochemistry, and neurology providing peeks into what consciousness is and how it works. Their findings give us intriguing clues as to what is actually happening in and through our bodies, minds, and spirits as we roll pen across paper. Each researcher has his or her finger on the pulse of a stream of information in a particular field. Taken together, these scientists are pointing to a new, provocative, and very compelling view of the cosmos and our place in it—a view that helps explain how and why writing has such a profound impact on our souls, our spirits, and our lives.

James W. Pennebaker, Ph.D., the chair of the psychology department at the University of Texas at Austin, stumbled upon some interesting information about the power of inhibition early in his career. In 1985, he surveyed two hundred people at a Dallas corporation on their health, the severity of traumas in their lives, and whether or not they had disclosed those traumas to anyone else. In *Opening Up: The Healing Power of Expressing Emotions*, Pennebaker describes the surprising results:

> Those with the most health problems had experienced at least one childhood trauma that they had not confided. Of the 200 respondents, the 65 people with an undisclosed childhood trauma were more likely to have been diagnosed with virtually every major and minor health problem: cancer, high blood pressure, ulcers, flu, headaches, even earaches. Oddly, it made no difference what the particular trauma had been. The only distinguishing feature was that the trauma had not been talked about to others.

From this survey, it appeared that the negative results of inhibition could be measured, and if so, then the positive results of disclosure should be measurable as well. To test this hypothesis, Pennebaker had three groups of college students write for ten minutes a day for four days. The first group was told to write a detailed description of a traumatic experience in their life—but only the details of what happened. The second group was told to write about their feelings about a traumatic experience—but only their feelings. The third group received instructions to write about *both* a traumatic experience and their deepest thoughts and feelings about what happened. The third group experienced lower blood pressure, higher immune function for six weeks after the study, improved mood, and better physical health for six months. Not only did the two other writing groups, the event-only and the feelings-only groups, not experience those positive results, but the feelings-only group also felt worse.

The results were so intriguing that Pennebaker began testing expressive writing in numerous circumstances. The only constant was that the students were asked to write for ten minutes a day for four days. Successive research projects demonstrated that students who wrote in an open, meaningful way developed a more positive outlook on life, experienced reduced anxiety and depression, had an improved ability to fight infection with more active T-lymphocytes (the agents in the body that fight cancer), and had lower heart rates. They also—this is my favorite—received better grades!

These results were not limited to students. Pennebaker tested the impact of writing on older adults as well. In 1991, an outplacement firm in Dallas asked Pennebaker to work with fifty male engineers, averaging fifty-two years of age, who had been terminated unexpectedly. In *Opening Up*, Pennebaker describes the study:

Even though they were among the most bitter and hostile group of adults I have ever seen, they were eager to try anything that might increase their odds of finding another job.

The basic study was quite simple. Half of the men were asked to write about their deepest thoughts and feelings about getting laid off for 30 minutes a day for 5 consecutive days. The other half wrote for the same period about . . . time management . . . . A third group of 22 men did not write at all and served as a comparison [or control] group.

As with our other studies, the men who were asked to write about their thoughts and feelings were extremely open and honest in their writing. Their essays described the humiliation and outrage of losing their jobs as well as more intimate themes—marital problems, illness and death, money, and fears about the future.

The potency of the study surprised even us. Within 3 months, 27% of the experimental participants landed jobs compared with less than 5% of the men in the time management and no-writing comparison groups. By months after writing, 53% of those who wrote about their thoughts and feelings had jobs, compared with only 18% of the men in the other conditions. Particularly striking about the study was that the men in all three conditions had all gone on exactly the same number of job interviews. The only difference was that those who had written about their feelings were offered jobs.

The results may have surprised Pennebaker and his students, but I understood immediately what had happened. From eleven years in the headhunting business, I knew that expressing anger at your previous employer is a major way people blow job interviews. "Why are you available?" is a standard interview question. Candidates have to give an accurate but unemotional answer, and they have to deliver it without a hint of hostility creeping into their voice. The angry emotions walled up inside these men needed someplace to go, but where? They couldn't express their full fury and fear with their families who were already scared about Dad's sudden unemployment. And they certainly couldn't express these strange and frightening emotions with safely employed friends, neighbors, or golfing buddies. So when an interviewer asked, "Why are you available," the non-writers' bottled up anger spewed out, ruining the interview. But the writers, who had already dumped their anger onto the page, were able to calmly answer the question.

Another scientist who is uncovering amazing new information that can help us understand what happens when we write is Candace Pert, Ph.D. You may have seen her in the movie *What the Bleep Do We Know?* or in Bill Moyer's PBS series, *Healing and the Mind*. Pert is famous for discovering that every cell of our body has receptors for peptides, which she calls "the molecules of emotion," and these peptides occur instantly and everywhere, communicating the information that runs all the systems of our bodies. Her research makes it clear that we are not a body with a separate mind, but a unified psychosomatic network that she calls the bodymind. In her first book, *Molecules of Emotion*, Pert describes the brain chemistry that supports and explains Pennebaker's findings on the damage done by inhibiting upsetting experiences and feelings:

> The brain's only food is glucose, which is carried to the brain in the blood. . . . Blood flow is closely regulated by emotional peptides, which signal receptors on blood vessel walls to constrict or dilate. . . . If our emotions are

blocked due to denial, repression, or trauma, then blood flow can become chronically constricted, depriving the frontal cortex, as well as other organs, of vital nourishment. This can leave you foggy and less alert, limited in your awareness and thus your ability to intervene into the conversation of your bodymind, to make decisions that change physiology or behavior. As a result, you may become stuck—unable to respond freshly to the world around you, repeating old patterns of behavior and feeling that are responses to an outdated knowledge base.

Unexpressed negative emotions keep us stuck in our old behaviors. That word *stuck* is all too familiar. If I could hear a recording of myself during my divorce and the years leading up to it, I'd hear myself saying over and over and over again, "I'm stuck" or "I feel stuck." What about you? If your angel handed you a recording of yourself complaining about the problems in your life, would you hear yourself repeating, "I'm stuck"? "Stuckness" seems to be a universal condition and one we all want to change. Pert has good news about how to change it:

> [R]eceptors are not stagnant, and can change in both sensitivity and in the arrangement they have with other proteins in the cell membrane. This means that even when we are "stuck" emotionally, fixated on a version of reality that does not serve us well, there is always a biochemical potential for change and growth.

But how do we actually change? How can Pert's science help us improve our lives? She has an answer:

By learning to bring your awareness to past experiences and conditioning—memories stored in the very receptors of your cells—you can release yourself from these blocks, this "stuckness." But if the blockages are of very long standing, you may need help in achieving such awareness, help that may come in many different forms. I would include among them psychological counseling, hypnotherapy, touch therapies, personal-growth seminars, meditation, and prayer.

By linking Pert's research to Pennebaker's, I think we can definitely add writing to that list. But how does writing release those blocked memories?

I asked Robert and Michelle Colt that question. The Colts have an unusual brain-based coaching practice. They help corporate executives, professional athletes, and actors and screenwriters use their natural brain functions to release the pull of the subconscious mind when it is in direct conflict with conscious desires. In other words, they teach people how to finally get what they say they want. The Colts are both certified master practitioners and master coaches in Neuro-Linguistic Programming (NLP) as well as certified master practitioners of hypnosis. But, as Robert told me in a telephone conversation in August 2007, "That's just the linguistic part; the key is that we have used our own brains as guinea pigs to discover how the brain really works." They worked with James Hardt, Ph.D., a physicist, psychologist, and psychophysiologist with over thirty years of research and clinical practice in neurofeedback brain-wave enhancement. At Hardt's Biocybernaut Institute, the Colts not only learned how the brain works, but they also *experienced* how the brain works.

"Candace Pert is right," Robert told me, "the brain is plastic. You have a lifelong ability to create new neural pathways. But to do that, you have to change your beliefs."

"Oh," I said, "you mean change your thoughts."

"No," he said (and this surprised me), "thoughts are powerless. They are neutral. It's the belief you have in them which has the power. It's all in the emotion. Think of emotion as energy in motion. But you have to remember that the brain, although capable of change, is geared first toward survival. When you have negative, fear-inducing experiences—and we all have them growing up—the brain creates neural pathways. And over time you become locked into those repeating patterns of experience because you stop looking for, or even being aware of, anything different. Something new could be out there—it undoubtedly is—but your brain is focused on attracting and repeating the experiences that it knows. And with each repeated experience confirming your negative emotions, those neural pathways get carved deeper and deeper until they are no longer pathways, they are more like neural highways. That's what we experience as habits, and habits, as we know, are difficult to break."

"So can writing help you break them?" I asked.

Michelle jumped in. "Absolutely. When you write, you use several modalities at once: visual—you see what's on the page, and you also see the events you are writing about in your mind; auditory—you hear yourself talking to yourself in your head, and you can actually amplify that by speaking out loud; and kinesthetic—you feel the pen, the paper, the whole physical experience of writing. That alone—using all three modalities—makes writing very, very powerful."

"But what exactly happens in your brain when you write?" I asked.

Robert explained, "First, you have to know that there are four types of brain waves. Beta, the fastest, are associated with stress, work, and concentration. When we awaken to an alarm clock and immediately start focusing on all the things we have to do, we are leaping from the slowest brain waves of sleep right into high-speed beta. Most of us live the bulk of our days in fast-paced, high-stress beta. Alpha waves are a bit slower than beta. They are connected with creativity, calmness, and insight. We are in alpha when we engage in good, almost effortless, work. You've heard people talk about being 'in the zone.' That's deep alpha. Theta waves are the next slowest. We all experience them when we awaken naturally. During those drowsy moments of theta, we remember our dreams vividly and can have truly creative ideas and breakthroughs. Almost everyone has had the experience of waking up knowing the solution to a problem. That's theta. Brain scans of people who meditate show that they drop quickly into deeper and deeper layers of theta and remain in theta while they meditate. The slowest brain waves are delta. We experience them when we are deep asleep."

"When someone writes, what brain waves are they experiencing?" I asked.

"When you first begin to write, you probably start out in beta mind, particularly if you are writing about a stressful event or trauma, like the people in Dr. Pennebaker's studies," Robert explained. "But very quickly, you will pop into some mid to deep alpha and eventually theta. Any moment of intense creativity is a theta burst. And when you engage in deep dialogue with divine mind, you are having mystical theta bursts."

"Whoa," I cried, "*mystical* theta! That sounds fabulous! How do you know when you're in mystical theta?"

Robert's accurate but unsatisfying answer was, "You know."

And he's right. You do know. I've had the experience—and you will, too—of seeing something on the page and knowing that it did not come from you. It came out of your hand and perhaps even out of your mind, but it did not come from your knowing, your beliefs, your desires, your current consciousness. It came from somewhere else, somewhere deeper. Eventually, as you write often enough to experience this regularly, you will begin to know when you are no longer the writer doing the writing. You will feel that something else is helping you move the pen. I know it sounds a bit wacky and perhaps even a little scary, but when it happens, it's not scary at all. It's delightful. You will whoop in joy as you realize that you are truly connected with the Voice.

"So when you write," I asked, "are you literally changing your brain?

"Yes," said Michelle, "one of the wonderful things is that writing opens you up to new learnings, new neural plasticity. Your brain does have the ability to learn, and as you write, you create new neural pathways. You are literally moving your brain. You are breaking old neural habits and creating new ones. It may be hard at first, but over time it gets easier."

Now here was something I wanted to know. "When does it get easier?" I asked.

"It takes thirty days to create new neural pathways," Michelle said. "When you write, dedicate thirty days in a row. If you break the cycle, you'll have to start all over again. We know through neural SPECT [a form of sophisticated biofeedback equipment] scans that the new pathways will recede if you don't maintain them, and you have to go back to the starting point all over again."

"Wait a minute," I said. "Is that why people can go to a wonderful workshop, or have some sort of spiritual epiphany, or read a book and swear it's changing their life, and then—boom—you see them a few months later and everything is exactly the same as it was— or worse? They have the same problems, the same fears, the same unhappy relationships. Is that because the old neural pathways 'won out' so to speak?"

"That's it," Michelle said. "What you see out there in your world is a movie, a reflection of what's being run inside. When you create new thoughts and beliefs supported by new emotions, and repeat them consistently to create new neural pathways, you get a new— and better—movie. We see it in our practice every day. People who are willing to step out of their comfort zone, tap into source energy, and use it to create new beliefs, dramatically change their lives."

Source energy—I wondered if that was a scientific term for divine consciousness or what I experience as the Voice.

"It's all energy," Robert said, "call it infinite intelligence, or God, or whatever you want. It is source energy—the most powerful and intelligent pattern that expresses itself through you or me. You are energy. I am energy. So we are always connected to source energy."

But Robert's answer still left a few questions: *How* do you connect with source energy? *How* do you create those new beliefs? *How* do you generate those new neural pathways? When your life isn't working the way you want and you feel frustrated, scared, lonely, and worried, it doesn't work very well to just *tell* yourself to think better thoughts or feel more positive emotions. We've all tried something like that under the guise of "positive thinking," and we've all experienced how poorly it works. Yet I know that writing accomplishes all this and more. I know that it connects you with source energy, helps you create new beliefs, and moves you to those new neural pathways. I know it from the dramatic changes writing has produced in my life and in so many other soul writers' lives. But I wanted to understand *how* it works, so I could break it into practical, easy-to-follow steps and pass them on to you. So I turned back to my favorite book on the power of writing, *Opening Up*, for some enlightening answers.

After seeing the dramatic benefits of expressive writing, Pennebaker wondered if all forms of self-expression would produce the same positive results. He tested two other forms of expression, singing and drawing, and found no measurable impact. Next he tried physical movement. Working with a doctoral student who was also a dance therapist, he set up a study to test the effect of movement. Sixty-four students were divided into three groups. The first expressed a traumatic experience through freeform dance for ten minutes a day for three consecutive days. The second group danced out their trauma as well, but—this is important—they also wrote for an additional ten minutes. The third group, used as the comparison or control group, just followed an exercise routine; they did not write or express any trauma in their movement. Not surprisingly, the control group exhibited no benefits. Both groups of dancers reported that they enjoyed the opportunity to move expressively, but "only the movement group that also wrote evidenced improvements in physical health and grade-point average." Pennebaker concluded that, "The mere expression of a trauma is not enough. Health gains appear to require translating experiences into language."

You have to use language—that's a pretty stunning conclusion. But what language? Will any words do? Do certain words produce the most beneficial effects? Is a particular way of writing more effective? To answer those questions, Martha Francis, a University of Texas graduate student working with Pennebaker, developed Linguistic Inquiry and Word Count (LIWC), a software program that evaluates the frequency of word usage in writing samples. She and Pennebaker ran writing samples from Pennebaker's previous studies through the LIWC software.

The results were clear. There were three types of words that appeared over and over again in the writing samples of the subjects who exhibited the most beneficial results. First, the subjects used positive emotions words, such as *love*, *happy*, or *good*. Second, they also used negative-emotion words like *angry*, *hurt*, or *ugly*, but they used them in moderation. Both of these findings are interesting and support the power of emotions so clearly articulated by the Colts and Pert. But the big "aha" came with the third result: writers who demonstrated causal thinking, insight, and reflection, expressed through words such as *understand*, *realize*, or *know*, had the greatest benefits. The researchers noticed something else. Not only did the people who experienced the greatest benefits use emotion and reflection words, but their writing samples also changed over time from disjointed descriptions to coherent stories with a beginning, middle, and end. "Writing," Pennebaker concluded, "moves us to a resolution."

Pennebaker also found evidence of what Michelle Colt called new neural pathways. He hooked up volunteers to brain wave measurement equipment as they wrote about a traumatic experience for one minute and then stopped and wrote about their plans for the day for one minute. They went back and forth, writing expressively and then writing something mundane. The equipment showed that while they were writing about their trauma, the brain-wave activity on the left and right sides of the brain was much more highly correlated than during the time they were writing about their day. Pennebaker concluded that confessional writing brought about "brain-wave congruence." Why was this congruence significant? Because, as he explains, "our conscious thought is highly dependent on the language capabilities provided in the left brain. The parts of the brain that control negative emotions tend to be localized on the right side of the brain." That means that writing and language connect the story in your left brain with the negative emotions in your right brain that keep you locked in your old neural pathways. As you write, the two sides start converging.

Robert Colt helped me understand why this convergence is a good thing. "The right and left brain is connected by the corpus colosum," he explained. "As you get more deeply involved in writing, entering a meditative state, the right and left hemispheres start coordinating into whole brain functioning, and you experience theta, possibly even mystical theta, waves."

Those mystical theta waves are, I see now, the doorway between conscious mind and cosmic mind, self energy and source energy, old neural pathways and new neural pathways, life as it is and life as it could be. When we enter that mystical doorway, we walk into a new place, a place we've never been before. And there, we can ride our brain waves from old thoughts to new thoughts, from negative emotions to positive emotions, from confusion to resolution. As our words converge on paper, we can see and hear the Voice, and when we follow the guidance it conveys, our world, or our "movie" as Michelle Colt calls it, can change.

I am grateful for the science and the scientists who are showing us what happens as we write. (If you, like me, are captivated by all this new science, I invite you to explore more in the Resources section.) But just knowing the science is not going to change your life. Using it is. And that's what the next section of this book is all about—the *how*: how to tell your story, how to activate the Voice, how to attract and recognize guidance, how to discern the Voice from all the other noise, how to work with your inner critics, and how to tell when writing is working.

How Do I Write Down My Soul?

Once we've heard about something special—whether it's the ultimate recipe for perfect baked chicken, the model's secret to toned inner thighs, or the investment strategies of the insiders—we all clamor for one thing: how. How can we—the ordinary, flabby, and not-so-rich—get in on the scoop.

So here it is—the scoop. And it applies to you. Because everyone has direct and immediate access to the Voice. *Everyone.* You just need to know how.

There are four steps to writing down your soul and luckily they're easy to remember:

1. Show up
2. Open up
3. Listen up
4. Follow up

It's an accident of language that each step ends in up. Or maybe it is not. Perhaps the implied direction in that spunky little preposition is the perfect metaphor. Starting in earliest childhood, when the adults around us spoke of godly things they invariably rolled their eyes skyward. And when they mentioned the divine, they pointed a finger "up there." Even the burliest, when they joked about "the Big Guy," pointed at the ceiling. I'm sure this habit began eons ago when our first ancestors struggled to comprehend the incomprehensible. Can't you just see them lying on the ground, looking up at the night sky, staggered by the uncountable specs of light, and concluding that up there, way up there, must be the home of the mysterious force. And heaven and all its denizens have been "up" ever since.

Even though the sainted ones from almost every spiritual tradition teach that the divine is within and the kingdom is at hand, our language still clings to the ancient idea of the divine up there, the kingdom up there. So even though where you're going is deep, deep within, linguistically everything is up. The way to write *down* is *up*, the way *into* your deepest knowing is *up*, the way *out* of your current situation is *up*.

Step One: Show Up

"Show up" sounds like a self-evident step—more of a "duh" than an instruction. But it isn't a duh at all. For many, it is the hardest and the bravest step, because it is the first step into the unknown. Showing up marks the first creak of the door into the realm within. Step one may seem self-evident, but the truth is, if you don't show up to have the conversation, the conversation can't take place. The

Voice is always ready, always available. The question is, are you ready and willing to have the conversation?

How exactly do you show up?

Set a Time and Place

Setting a time and place sounds so simple, so straightforward, so matter of fact. But it's not simple at all. Our crazy lives are already overfilled with work and family and relationships and children and money and school and cars and shopping and cooking and cleaning and sleeping. There's already not enough time for all those obligations, so how do you squeeze another fifteen minutes into your day?

Here's the answer: you decide what is important and what can wait for fifteen minutes. You decide if you want to speak with the Voice or not. You decide if you need or want the guidance.

It's your choice—at least for now. But if, like me, you decide that something else is more important, the universe may do the deciding for you. And if the universe needs to get your attention, the universe *will* get your attention. Sometimes the attention-getting device takes the form of you getting fired, sick, divorced, dumped, rejected, broke, or run over. And in the moment these things are happening it's pretty hard to be grateful. It's only after you've survived the cosmic two-by-four and realized it was a blessing, that you find yourself saying, "Hey, if that bad thing hadn't happened to me, I'd never have fallen to my knees, and if I'd never fallen to my knees, I'd never have asked the Voice for help, and if I'd never asked for help, I could not be where I am today."

So I ask you: now or later? Do you want to have the conversation with the Voice now, before things are truly dreadful, or do you want to wait?

Sit Down

Once you've set a time and place, sit down in your chair at your chosen time.

The toughest part about showing up is the sit down part. It seems that the moment you declare your intention to engage in real spiritual practice, the world and all the people in it start coming up with new and sometimes quite novel things you should do instead. The phone rings. Your computer beeps. The cell phone rings. The fax spews. Your spouse yells. The washing machine lurches across the laundry room. The kids argue. You realize you haven't defrosted dinner. The bulb on the lamp snaps out. You notice a picture is crooked. You have to go to the bathroom. Your tummy hurts. You're suddenly thirsty or tired or anxious. *Something* happens to stop you from sitting down.

It is completely normal for the world to do everything in its power to get you to not talk with the Voice. On some level, you know and your brain knows and your people know that once you start talking with the Voice, things are going to change. And you and your brain and your people do not really think change is such a good idea, because no one knows how it's going to come out. And all concerned are totally uncomfortable with all this unknowing.

Recognize your interruptions for what they are: your little spirit's frightened efforts to keep Big Spirit from shaking things up. Yes, it's you creating the chaos. Not consciously, perhaps, but on another level, you are allowing and showing up for exactly what you say you don't want.

When you realize you are constantly being interrupted or interrupting yourself, take your questions to paper and ask the Voice to help you understand what is happening and what you can do to create the positive, focused, quiet writing time you want. These questions make exquisite early writing material.

WRITE DOWN YOUR SOUL

Dear Voice,
What is this? Why is it happening at this precise time?
Who is interrupting me? Where am I in this
distraction? In what way am I allowing it to happen?
What can I do—what do I choose to do—
to protect my writing time?

When I write, I do not answer the phone. I've decided that this tiny block of time in my day is for me and the Voice; everyone else can wait. And guess what? Once I became clear about this boundary, the world stopped trying to break it.

Loretta came up with a brilliant solution to the inevitable distractions that come with children. Every evening, she told her nine-year old daughter, "Mommy needs this time alone for just ten minutes." Then she would set her daughter up outside the bedroom with a pile of toys and books. And every evening, her daughter would find some reason to come in and interrupt her.

One night, instead of begging her daughter to leave her alone, Loretta patted the bed beside her and said, "Honey, why don't you come up here and write beside me?" Ripping a few pieces of paper from her journal, she said, "Write a letter to God, Baby. God wants to hear from you." Her daughter put her head down and started scribbling intensely. Loretta was able to write for twenty minutes or more. Writing time became the sweetest part of the day for both of them.

One evening her daughter showed Loretta a picture she'd drawn and said, "This is Dad when he's drinking and I hate him. I drew this for God in case God doesn't know." For a second, Loretta thought, "Uh oh, maybe this isn't such a good idea," but then she realized how much her own writing was helping her deal with exactly the same problem, so she smiled at her daughter and said, "It's a lovely picture, sweetheart. I'm sure God really likes it."

Write at Your Chosen Time and Place
Every Day for Thirty Days

You have the potential to communicate with the Voice, to experience those mystical theta bursts. But just the desire to do so is not sufficient; it takes a commitment to show up for your spiritual exercise. Just as your body looks better and feels stronger when you exercise regularly, your spirit will look brighter and feel much stronger as you build your spiritual muscles.

So sit down the first day and the second day and repeat for a total of thirty days. This spiritual regimen may feel weird or uncomfortable at first, but, just like a new workout routine, over time it gets easier and your confidence grows in your ability to do it and do it well.

At every Writing Down Your Soul workshop, attendees ask me if they really have to write every day for *thirty* days. Not long ago, the standard axiom was that it took twenty-one days to undo a habit or create a new one. That's what most performance consultants and spiritual teachers said, so that's what I taught, too. I also used to say that if you couldn't write every day, at least write five days out of seven. Sure, I had written every day during my trauma, but I wasn't sure that others could make that commitment, and I didn't want to scare anyone away from such a profound practice by imposing what seemed like an arbitrary rule. But invariably, when the class met twenty-one days later, it was clear that some writers had developed a powerful spiritual practice and others had not. This pattern frustrated me. I couldn't understand why three weeks seemed to be plenty of time for some and completely inadequate for others, especially those who didn't write every day.

When I heard the Colts explain that there is actual physical construction going on inside our brains when we write, I understood why I had to stick to the "every day for thirty days" recommendation. Those new neural pathways don't open up by themselves; it takes real effort over real time to open them and then hold them open. So mark off thirty days on your calendar, then write every day, even if only briefly on some days. Better to write for a few minutes than not at all.

Write at the Same Time

New writers also ask me if they really have to write at the *same* time every day. I'm always torn when I answer. I'd love to tell them to write when they can. But I know from experience that people who don't create a pattern don't write consistently. And people who don't write consistently have difficulty opening that wisdom door and keeping it open. Then, when they do write, they can't be certain if they're hearing the Voice or doing all the talking themselves. Consequently, change for them is limited and sporadic.

Get Ready: Create and Use
Your Personal Writing Ritual

When you show up in your writing place at your writing time, with a journal in your lap and a pen in your hand, get ready to have the most profound conversation of your life.

How do you get ready?

Breathe Deeply a Few Times

Take a deep cleansing breath. As you are breathing in, visualize all the crud in your system flowing out of your cells and into your center. Hold your breath for several seconds while your lungs fill up literally and metaphorically with everything you want to release. Then breathe out for a slow count of three or four. As you breathe out, visualize all that cellular junk—physical, mental, spiritual, and emotional—leaving your body. Do this type of breath two or three times. This kind of deliberate breathing slows your heart rate, brings oxygen deep into your lungs, and helps you become centered in the present.

It seems that all spiritual and physio-spiritual practices such as yoga begin with deep cleansing breaths holding the breath a bit before exhaling. I never thought much of it when the minister suggested we take a few deep breaths before meditating or when my Pilates trainer told us to breathe deeply for a count of eight on the inhale and exhale. I just followed instructions. But then I read about the importance of breath in James Hardt's *The Art of Smart Thinking*:

> An ample flow of blood to oxygenate the brain is critical to the brain's ability to create alpha waves. . . . [W]hen you shut off the supply of fresh oxygen to the body [by holding your

breath], the carbon dioxide content of the blood starts to increase. The body's automatic response is to expand the carotid arteries that feed your brain, which then allow more blood to flow to the brain. If this is practiced regularly, the carotid arteries expand permanently, and your brain is constantly provided more oxygen. We know that an oxygen-rich brain is necessary to produce alpha waves.

Since reading Hardt's book, I have renewed appreciation for the cleansing power of the breath, and I breathe deeply when I sit down to write.

✳ Set Your Intention

Intention is critical. It gets everything in motion. A short prewriting prayer or brief blessing sets your intention to access the Voice and reminds you that you are communicating with something holy. Raise your writing hand and say something like this:

> Here I am. Take my hand. I am ready, willing, and worthy to speak with you right here and right now. Thank you in advance for your words, your wisdom, your guidance, and your grace. Amen.

Did you stumble over any part of that blessing? If you did, I'll bet it was "and worthy." Many of us grew up in religious traditions or family systems that did not tell us we were worthy of anything, never mind the complete, unconditional love of the divine. If you find that phrase uncomfortable to say, say it anyway. Say it precisely because it is hard to say. It is the truth. You are most certainly worthy to connect with the divine and hear the guidance of Spirit, and even if no one else knows it, I know it. So say it.

If you want to maximize the power of this little prayer, say it out loud. If you feel silly doing that, just sit still for a moment and imagine the pure light of source energy pouring down on your head and hands. Some people visualize white light, others pale blue, some gold, some pink. Pick a color, or just see for yourself what color it is as it flows through you.

Another way to "say" your intention-setting blessing is to visualize the wisdom of the universe entering your third eye. Your third eye is in the middle of your forehead, right in front of your frontal cortex. The frontal cortex, as we've seen, produces language, chooses behavior, and processes emotions—all integral elements of writing down your soul. Your third eye is a conduit between you and unlimited knowing. To remind yourself that by writing you are learning to "see" in a new and deeper way, touch your writing hand to your third eye and know that you are creating a connection between your hand and the wisdom of the universe.

Whether you say a blessing, visualize the flow of pure light, touch your third eye, or invent a new method altogether, you are consciously knocking on that cosmic door and saying in some way, "Here I am. I'm ready. Come talk with me."

"Here I am." That was prayer enough for Moses. He saw a bush on fire and wondered why it wasn't turning to ash, so he stepped closer.

> When Yahweh saw Moses going across to look, Yahweh called to him from the middle of the bush, "Moses, Moses!" "Here I am," he answered. (Exodus 3:4)

The first time I heard Ron Roth, author of *The Healing Path of Prayer*, speak, I was deeply moved by his incredibly simple prayer, "Come, Holy Spirit, come." When Ron calls on Spirit to enter a place, you feel Spirit moving. If you like the simplicity and power of Ron's prayer, use it. Or simplify it to, "Come, Spirit, come," "Come, Voice, come," or just, "Come." If you want to amp up the power of your invocation, combine Roth's call and Moses' response, "Come. Here I am."

Read a Bit

During my dark time, my life was so jumbled, my problems so vast, and my mind so scattered, that I often didn't know what to write about first. I discovered that a bit of reading helped me get focused. Often the day's short readings matched my state of mind and helped me dig beneath the surface of my immediate problem to a deeper underlying issue or question. I offer you my reading ritual not as a formula to copy, but as an example of how reading can support and inform your deep soul writing.

I started by reading the short passage for the day in the little inspirational monthly magazine *Daily Word*. Millions of people read the *Daily Word*. It is written in gentle, positive, non-denominational language. At the bottom of each page is a short Bible quotation that inspired the daily passage. Sometimes I looked up the quotation in my battered New Jerusalem Bible. If the quotation came from the New Testament, I'd also look it up in *The Five Gospels: What Did Jesus Really Say?* This book is the summation of six years of work by a panel of scholars from numerous fields who met regularly to share their expertise and to debate the likelihood that the historical Jesus actually said something attributed to him in today's versions of the gospels.

After I finished digging around in my sacred texts, I'd switch to a bit of secular reading. Over the years, I read snippets from many books, but the one that carried me when I was hurting the most was Sarah Ban Breathnach's *Simple Abundance*. As its subtitle says, the book is a "Daybook of Comfort and Joy." Comfort and joy were two things I did not have and two things I wanted desperately, so this book was perfect for me. Many times, I found exactly what I needed to hear in the day's passage.

Often, an idea would grab me as I was reading. Hearing a question form in my head, I would throw down the books, grab my journal, and start writing.

If you want to add a short reading period to your spiritual practice, select your favorite sacred texts and put them next to your writing chair. Choose the works you love or something new if you feel a desire to explore. And don't limit yourself to traditional sacred texts. You can read any spiritual or, for that matter, secular material that speaks to you. Reading is simply another way to let the wisdom of the universe reach you. Experiment until you find material that works for you.

Your prewriting reading can last five minutes or so; longer if you are really enjoying what you're reading and getting a lot from it. Just don't forget to write. Reading is meant to inform your writing, not substitute for it.

Record the Date

Open your journal and write the date at the top of a fresh page. This instruction may sound silly; after all, the Voice knows what day it is. But I recommend it, because if you write for years, and most people who begin this practice do, at some point in the future you may want to revisit your journals. Perhaps you'll want to remember a story, track some changes in your life, look up a detail you've forgotten, or remind yourself of some wisdom you received. Maybe you'll just want to stare at a page where the Voice spoke to you. Whatever the reason, you'll have a tough time finding the entry without dates. Now that I write about the spiritual power of writing, the dates in my journals make it possible for me to share real world examples of that power with you. You don't have to write the date, of course, but for me it's part of the ritual.

Write a Salutation

Start with *Dear*, as in "Dear Voice," "Dear Spirit," "Dear One," "Dear Source," "Dear Friend." You choose the name, but do use the sweet term *dear*, because you are writing a letter. We rarely write letters by hand today, but when we did, we always started "Dear____," and that still seems like the best way to begin an intimate written conversation.

You could, I suppose, say *Hello*, but that seems so ordinary and this salutation is not the beginning of an ordinary conversation. Speaking directly with the wisdom of the universe deserves something special.

Create Your Own Writing Ritual

I've shared how I show up to communicate with Spirit. This ritual works for me, but by no means is it the only way to establish a writing practice. Just as you get to decide how you want to address the Voice, you get to decide when and where and how you want to write.

A year after taking a Writing Down Your Soul workshop, Sylvia, a teacher, wife, and mother of four talked with me about the writing ritual she created:

I get up every morning at five and write for ten minutes. I go right into my study with a glass of water and light a candle. I don't read before I write. I read after. I read a few lines in the lessons in the back of the student workbook for *A Course in Miracles*. It seems that my writing and my short reading in the lessons are somehow always connected. Often, the reading seems like an answer to what I wrote.

At first, my husband would constantly come in and say, "What are you doing?" in a sweet, teasing kind of way, but still, it was an interruption. Finally I told him, "Honey, I really need this time for myself," and he stopped interrupting me. Now I have this protected, precious writing time just for me, and I write almost every day.

Use All Your Senses

I was excited to learn from Michelle Colt that writing is effective because it utilizes three senses: sight, hearing, and touch. But after talking with my friend Sylvia, I realized that by using the candle and water, she had also incorporated the last two senses, smell and taste, into her deep soul writing. One spiritual practice utilizing all five of our physical senses!

Since meeting with Sylvia, I've started lighting a scented candle and putting a glass of water next to me. But I don't drink the water while I'm writing. I am so engrossed and writing so fast, that I don't want to break my focus to stop and drink. Instead, I use the water in the way I learned in a Navajo ceremony with a blind medicine man in 2006. After a long ceremony to remove a personal energy block, I was handed a glass of water. The medicine man explained that in a Navajo ceremony, all the divine energy and grace that is drawn to you during the ceremony is held in the water. He instructed me to drink it mindfully, slowly, and fully, visualizing the healing energy in the water entering my body and soothing every cell.

Thanks to Japanese researcher Masaru Emoto's groundbreaking experiments, we have scientific evidence that water is transformed structurally by our thoughts and words. In astonishing photographs in *The Hidden Messages in Water*, Emoto showed that water that is blessed and praised transforms into beautiful snowflake-like cellular structures, and water that is cursed or exposed to harsh words or music distorts into non-symmetrical, ugly forms. I figure my glass of water, sitting a foot away from my soul journal, is exposed directly to the energy of my conversation with the Voice, so it must be beautiful water indeed.

Now I drink a glass of water slowly and consciously after I finish writing. I also swallow my daily vitamins while drinking my blessed water; I look at each vitamin and say out loud something I learned during my writing time. I am confident that swallowing the water and the vitamins in this way solidifies and strengthens my writing experience.

Tom Nicoli, an internationally certified, multiple-award winning, board member of the National Guild of Hypnotists, told me, "Until it's written, it isn't real. A thought is just the beginning of the process of making that thought a reality. By writing our thoughts we make them more tangible and include the physical and visual components which make thought more powerful and attainable." So record your soul's choices for your writing practice in this chart. Don't skip this part of showing up! Fill in your answers and make your writing ritual real.

My Writing Down My Soul Practice

My writing place: _____

My writing time:_____

My statement of intention (e.g., prayer or blessing): _____

_____

_____

My prewriting reading material: _____

_____

My chosen salutation: _____

My opening writing ritual: _____

_____

Step Two: Open Up

There you are in your writing chair. You've taken a deep breath, set your intention, done a bit of reading perhaps, and now it's time to write. Many people find that as they add the comma at the end of the salutation, the words rush forward to be heard.

Elizabeth, an avant-garde artist, wanted to start writing down her soul, but she was preparing for an art opening and could not come to a workshop. So I gave her the world's shortest lesson on how to write down your soul: address the Voice directly, write fast, ask lots of questions, and write whatever comes. Two weeks later, she called to say she was amazed by the thoughts flowing unbidden out of her pen—thoughts she had not realized were inside her. As Elizabeth and all deep soul writers discover, there's something about speaking directly and intentionally to the Voice that frees the heart and mind to open and release all that is stored inside.

And that, in a nutshell, is the key to this step: *open* your heart and your mind—all the way—and just say what wants to be said.

Look at the first half of the Möbius strip that depicts the communication highway between you and the Voice.

The first half says, "I write. The Voice listens." In other words, you go first. The Voice will speak, but first, you have to tell your story. But you may not tell it the way you tell it to a close friend, to a spouse or lover, or to a spiritual advisor or therapist. When we talk with people, there is always an element of judgment, whether it is acknowledged or not. We want to connect. We want to be heard. The other person wants to connect, too. But the other can't help but bring his or her feelings, experiences, and wounds to the conversation. In *Writing to Heal*, James Pennebaker captures this problem:

Talking to someone about a trauma is far more complex than writing about it. To the degree the other person accepts you no matter what you say, and you can be completely honest in your disclosure, then talking may be more effective than writing. But here's the rub. If the person you confide in does not react favorably to you and to what you have to say, then talking may actually be worse for you than not confiding at all.

We've all had this painful experience. Often it happens early in a new relationship when everything feels good and fresh and oh so safe. You open your mouth, thinking you've finally found someone who wants to embrace the total you, but when you finish, the reaction you hoped for isn't there. Instead of smiling knowingly, nodding enthusiastically, hugging you warmly, and declaring, "I know exactly what you mean," the person looks at you oddly, or says you're wrong, or starts telling you how to fix your problem, or—worst of all—doesn't say a thing. Whatever your listener does or doesn't do, you clearly get the message that you've crossed a line and now you have some serious backpedaling to do if you want to salvage the relationship.

Once burned by the repercussions of open-wide communication, you step warily into your next relationship, holding your deepest feelings and secrets in abeyance until you're sure it's safe to let them come out. And even then, you move slowly and cautiously because you've learned that it's best to give away little doses of your true self and wait to get a reaction before disclosing any hint of the abyss within. Yet the abyss is there. The confusion is there. The pain is there. But the people around you are rarely, if ever, totally accepting of your full, rich, complex, emotional self.

Why? Why are we so uncomfortable with full-open, honest communication? Because society has all kinds of norms around connection and communication, and we are all raised to learn quickly what is acceptable and what is not. After a few sharp reprimands, we learn that we can't skip down the street blurting out, "She's fat!" or "His clothes are dirty." And we learn to stop expressing the storms bubbling in our hearts. We discover that screaming "I wish you were dead" or "I hate you" will get us sent to our room—or worse. We learn that crying from the depths of our souls just frightens people, and if they get frightened enough, they shout devastating things like, "Shut up, or I'll give you something to cry about." We also learn that outside, and sometimes even inside the family circle, people don't know what to do when we throw ourselves at them with intense affection. Anger gets squelched. Fury is stamped out. Adoration is unseemly. Deep pain frightens people. Even ecstatic joy can drive people away. Intense emotions just do not fit very well into our daily discourse.

But they fit perfectly into our conversations with the Voice, because the Voice does not judge. The Voice does not try to massage our emotions back into acceptable form. The Voice is not the arbiter of socially acceptable behavior. The Voice simply listens. And there is no greater gift than that. It's what we all seek—someone to listen, really listen.

Very few people know how to listen, because real listening is not taught anywhere. It is not modeled in the home, encouraged in the classroom, or covered in the college curriculum. So it should come as no surprise that failure to listen is a primary issue in couples counseling. We know that listening is essential to effective relationships, but we can't seem to do it. Too many beliefs and judgments get in the way. When someone tells us their story and gets to a part that makes us uncomfortable, all the things we think they should be doing jump up and down inside our heads. All that internal noise prevents us from really hearing what the other is saying. People can sense this, and they stop talking, or start avoiding certain topics, or start saying only what they think the other person wants to hear. Despite our deep human need to connect, we stop speaking our truth because the price for expressing ourselves fully and completely is too often the reduction or removal of love or the heartbreaking experience of watching it slip through our fingers before it's even begun.

In *Opening Up*, Pennebaker describes the ideal relationship in which you could safely disclose yourself. From the beginning, you must trust the listener and know without question that you are safe telling your story and expressing your deepest feelings. As you speak, the listener does not judge, recriminate, or criticize. In a footnote, Pennebaker says, "Ideally, if a person can find an all-accepting listener, talking to that person may be better than writing. The trick will be to find that listener."

The "all-accepting listener." That's a perfect description of the Voice. And that perfect listener awaits us every time we pick up our pen. We have only to open up, and we are heard.

Begin With What's Bothering You
or Happening to You Right Now

Sometimes we know instinctively and effortlessly where to begin writing. Other times, being given permission to speak freely to the perfect listener leaves us wondering: What *is* on my mind? What *is* happening? What *is* the most important thing to talk about right now? If you aren't certain, scan this list of questions. When something stabs your heart or lands like a rock in your stomach, take that question as your starting point. If none of these questions is exactly right for you, reading the list will trigger one that is.

- What am I worried about?
- What problem, if I could solve it right now, would change my life?
- If a wise therapist asked, "Why have you come to see me?" what would I say?
- What is my greatest fear?
- What sends my stomach into knots?
- What relationship is bleeding?
- What do I want?

Get out your journal, address the Voice, and record your question.

WRITE DOWN YOUR SOUL

Dear Voice,
My first question is: . . .

Write What Comes

Next, write the first thought that comes to mind—the very first thought—no matter how stupid, petty, blunt, or shameful you think it is. Just write it down. Do not think. Do not edit. Do not pause. Just write. First thought is a precious gift. It is an impetus from deep within your soul that wants to be seen and heard. Honor it. Writing it down doesn't automatically mean that it is true or has to happen. But it does mean that this thought is inside you and longs for expression. Let it out on to the page. Maybe your first sentence will come out something like this:

I'm worried about money.
My wife and I are drifting apart.
I think I'm going to lose my job.
Everyone but me seems to be successful.
I'm sick of arguing with my daughter.
My life isn't working.
I am afraid the cancer will come back.
I am lonely.

Explore Your Deepest Thoughts and Feelings

Whatever your first question and first statement, you have opened the door a slit. Crack it further by following Pennebaker's instructions: write about your deepest thoughts and feelings about this first statement. "Your deepest thoughts and feelings" means the whole story and your part in it. Don't leave something out because you think it's too awful or too scary or too mundane or too obvious or too pathetic to share with the Voice, because—I hope this doesn't come as a shock—the Voice already knows. And because the Voice already knows, you are safe to tell the truth and, at the same time, obligated to tell the truth.

Tell the Truth

This may be the first time you've been invited to open up and tell the truth, the whole truth, and nothing but the truth. It's an exciting invitation, but one you may find uncomfortable. It isn't that you don't want to tell the truth. You've come to this writing practice consciously or subconsciously because you want to speak openly and honestly about who you are, what you really feel, and what you really want. But for most of us, this kind of openness does not come naturally or easily. It feels strange to suddenly be given permission to open the valves all the way and let a free torrent of words pour out.

We've all had way too many years of carefully managing what we say in our human relationships. But try to manage a relationship with the divine? You can try, but it just won't work. At some point, you'll be writing about something that happened, and you'll get uncomfortable or feel ashamed. You'll gloss over a telling detail or not cop to the raw truth of your feelings, and suddenly your hand will write, "Well, that's not exactly true." Or at least it will if you are slamming down your thoughts as quickly as they come.

Write Fast and Ignore Writing Rules

Speed in deep soul writing is of the essence. This isn't a conversation with a parent who might disapprove or an adult who could get scandalized. This is a conversation with the all-knowing, all-loving, all-listening energy that created you and loves you and will not—cannot—remove its love. So you don't have to think twice before you write. At long last, it is safe to share your secrets. But because you've been conditioned for so long by the land mines of human communication, it may take a while for you to fully accept that safety. That's why speed is important. It's the only way to stay ahead of your internal editor. And oh, have you got an internal editor! Your internal editor is thoroughly steeped in the norms of your environment. For decades it has protected you from overstepping the bounds of what is acceptable in your family, your circle, your neighborhood, your community, and your culture. When you approach a socially unacceptable boundary in a human conversation, your internal editor pulls you back, whispering, "Don't say that."

You are so accustomed to following your inner editor's directives that you are probably oblivious to the sound of its voice. But when you pick up a pen with the conscious intention of accessing the Big Voice, your editor will go positively apoplectic trying to get your attention. You won't have to listen very hard to hear that little banshee screaming, "Don't admit *that!*" or "You can't talk to Spirit that way!" or "Shut up! For heaven's sake, *shut up!*" But you must not shut up. This is the one time and the one place in which you must strip yourself open and let your heart bleed all over the page.

When I first started writing down my soul, I was so angry that *writing* doesn't really capture what I was doing. *Vomiting* would be more accurate. Angry words gushed out of me onto the page, filling it with savage black marks. I screamed at God: "Damn it, where *are* you? Do you see *anything* that's happening here? What are you doing that is so important that you can't take care of us!" I swore at God. Yes, even the F word. The day I threw that on the page I felt awful. My internal editor was screeching mad, and all the well-trained little Catholic girl cells in my body were vibrating with guilt. I thought surely God must be angry with me.

I reached out for help. I called LaRee Ewers, creator of *Making Tired Eyes Smile*, a language-arts system for seniors with Alzheimers. I knew LaRee met regularly with a gifted spiritual director, so I was pretty sure she would have something helpful to say about the mess I was in. I told her what I'd done, expecting some advice on how to have a more acceptable dialogue with the divine. Instead, LaRee said, "What's the problem? Do you think God isn't big enough? God is big. God can take it."

Intellectually, theologically, in so many ways, she was right. But being right isn't what mattered to me. I had confessed, and LaRee's words had given me absolution. I was free. For the first time ever, I was free to be seen. I was free to let God see all of me, even—or especially—the blackest, angriest, scariest parts. I felt ten pounds lighter. The next morning, I grabbed my pen and dove into my journal with a vengeance. I was still angry, but now I was free to communicate my anger. And, oh, did I communicate it! I didn't care what my little goody-two-shoes internal editor wanted me to say; I knew I had major problems and only the truth could get me through them.

The best way to push your editor aside is to write fast—really, really fast. Make sure you have a pen that glides effortlessly over the paper. Don't bother spelling out names or places you mention over and over; just an initial or symbol will do. Don't bother with endings or punctuation or quotation marks or anything that slows you down. Don't worry about writing correctly—just focus on having a conversation. Because that's what deep soul writing is—a conversation. It happens to take place in a written form, but it doesn't have to meet any of the criteria of teacher-approved "good" writing. No one sees what you have written but you and the Voice, and the Voice doesn't care about your grammar, punctuation, or spelling. The Voice doesn't care if you never write a whole sentence. The Voice doesn't care if you leap madly from one idea to another. The Voice doesn't care if your writing is legible. The Voice doesn't care if your lines stay straight or zigzag all over the page. The Voice doesn't care if you stay in the margins or devour the whole page. The Voice doesn't even care if your thoughts make a lick of sense.

For example, you might be telling the Voice about getting annoyed with your daughter because she let a cup of hot tea that you made especially for her get cold and she poured it down the drain without taking so much as one lousy sip. And then suddenly you might be describing a scene in second grade when you didn't want to eat your asparagus, and your father spanked you, and you were shocked and mortified and angry and mad. And then you might leap to a vehement declaration of how much you hate your ex, and then you might dive into a list of unpaid bills, and then you might wonder why you're even doing this stupid writing thing because clearly the Voice is not listening and couldn't care less, and then you might start talking about a nasty problem at work and all the lunatics there, and then end with a giant plea for help.

The day's entry would be labeled a neurotic mess if a human read it, but it's a lyric poem in the eyes of the Voice. In that same scenario, if you wrote slowly and carefully in well-formed sentences—giving your internal editor plenty of room to "help"—everything would appear on the page clean, pretty, and thoroughly scripted. You might start off telling the same tea story, but then there would be apologies for your behavior, promises to do better next time, and all kinds of obfuscation of what really happened and why it all hurt so much. And there would not be a withering comment about the divine's refusal to listen—at least, not if you were raised in a tightly wound, theistic environment.

Try it yourself to experience the difference. Write slowly and carefully one day, thinking before you write, writing in full sentences, putting in the right commas and capitals, and going back to cross out words and make corrections when you think of something better to say. Create an entry that would earn an English teacher's approval. The next day (or right then, if you have time) write about the same experience, but write with abandon, jotting down phrases as if you were talking to a very special friend who wants to hear every word. Don't go back and edit anything. Just go forward, forward, forward, full speed ahead. You'll find there's a remarkable difference in appearance and content.

Joy, a young aesthetician with a broken heart, reported back to me on this experiment:

> I have gorgeous handwriting. I'm proud of my handwriting. So when you said to write fast, I didn't like it. My handwriting was no longer pretty. It actually hurt me to see mistakes and not go back and correct them. You even showed up on the page! I was frustrated and suddenly I wrote, "Janet says to keep going, so I'm writing fast. I don't like it, but I'm doing it!" My handwriting got huge. Huge! I'm shocked by the size of the words. Sometimes there are only four lines on the whole page! But since I started writing this way, what I'm writing about has changed. I used to beg Spirit to bring him back to me. Every night and every morning: "Bring him back to me. Why doesn't he come back to me?" But now, I'm asking for something different. Now I write, "I'm supposed to learn something here, and I don't know what it is. I'm ready. I want to learn. Show me!"

Joy's answers haven't come yet, but they will. The Voice cannot resist an invitation like that. Your job in Open Up is simply to release the words that want to be heard. If they seem stuck, issue them an invitation. Bless them. Tell them it's safe to come out. Tell them you'll keep the editor out of their way. Tell them the Voice is waiting to hear them because they are your story. And for the healing to begin, your story must be told.

Tell Your Story

Brian and Lisa Berman have a unique, intimate knowledge of the healing power of telling your story. They are facilitators for the Compassionate Listening Project, a nonprofit organization that teaches heart-based speaking and reflective listening skills to people in conflict around the globe. Brian and Lisa lead Compassionate Listening workshops in Germany that bring Holocaust survivors and their families together with SS officers and their families to listen to one another's stories. I was confident that Brian and Lisa's experiences with people telling their stories to one another would have applications for writers telling their stories to the Voice.

"Your story," Lisa began, "is your healer. In every story is conflict, and within the conflict is the chance for change, for growth, for development. The story is what happened; it is what is. The value is in what the story is telling you. It's your guidance from Spirit, who wants to bring you to a new place."

This description sounds like a direct parallel to writing down your soul. You have a story to tell, and your story, like everyone's story, is rife with conflict. But telling your story, especially if it's a deeply buried wound, can be difficult to do. I asked Brian and Lisa how they teach people to tell their stories.

"Safety is first," Lisa said, "There has to be a safe container." In Compassionate Listening, the safe container is the other members of the workshop, all of whom are trained to listen beyond the details of the story for the core identity and needs of the person speaking. When you write, your safe container is the Voice. The

Voice is always tuned into your core identity, who you really are, your very soul. When you sit down in your personal writing space, pick up your pen, and write with the intention of connecting with the unconditionally loving, all-accepting listener, you are in the ultimate container of safety.

Speak From the Heart

In your place of safety, you are free to speak as you have never spoken before—from your heart. Speaking from the heart, the Bermans explained, is different from speaking from the head. Typically, when we talk with other people we have an agenda, even if we aren't conscious of it. As we speak, we scan the other person for reactions. We look for clues that their experiences are resonating with our experiences, their wounds with our wounds, their feelings with our feelings.

Speaking from the heart is different. It is not about linking wound to wound. It is not about manipulating a response or gauging the other person's level of agreement. It is about uncorking and releasing the story from your deepest heart into the open, where, at long last, it can be fully heard and finally healed. Brian described how that happens. "As the people around the speaker listen reflectively to the story, the speaker is able to hear their own story, perhaps for the first time. Until then, they really hadn't listened to themselves. They kept saying the same thing over and over again and having the same angry feelings about it, but they never really heard what they were saying."

When you write down your soul, you tell your story—a story you have repeated many times before—but this time that story is different. This time you are not trying to trigger a sympathetic, wound-to-wound reaction. Because the truth is, you can't. The

Voice is not in the business of confirming our convictions or doling out revenge on the people we're upset with; the Voice is in the business of wisdom.

But what if there is a story you *don't* want to tell—a story you haven't told anyone, not even your best friend or spouse. What if you have a story you don't even tell yourself?

Emily had this kind of story. After college, she got a great job in Chicago. She rented a tiny apartment and spent her first few paychecks fixing it up. She loved her job and had a ball hanging out with her new friends at all the "in" drinking holes. Eight months later, she abruptly quit and moved back home. Her parents welcomed her with open arms, but when they asked what happened, Emily just said big-city living wasn't for her. They thought it odd, but soon Emily had a new job and an active social life and appeared to be back in the swing in her hometown. Seven years later, she went to a therapist because she couldn't shake off a bout of depression. During her third visit, Emily suddenly started talking about a date rape in Chicago. When she finished, Emily looked at her therapist in shock and said, "I'd completely forgotten this happened." Emily started writing down her soul that night. The more she wrote, the more she learned about her buried story and how it was impacting her life.

We all have buried stories. The night after Emily told me her story, I dreamed about a time when, as a college student in Milwaukee, I was riding the bus. An obese guy wouldn't budge from his seat to let me pass by when we reached my stop. In my dream, I stepped over him, and as I did, I felt him reach under my coat and rub my underpants. I woke in disgust—and surprise. I'd completely forgotten this incident, but more importantly, I'd forgotten how guilty I felt for not smacking him and yelling what he was doing.

Instead, I slunk out the back door berating myself for being so stupid. *That* was the story I needed to share with the Voice. I attacked it full bore in my journal. I asked why it bothered me so much. I asked why I'd buried it. I asked why it was important for me to remember it now. And I asked for help healing it.

In his research, Pennebaker discovered that people are least likely to disclose three traumas: parental divorce, sexual abuse, and violence. Your secret may be one of those, but it doesn't have to be. Throughout my divorce, I wrote with open veins, letting my emotional and spiritual blood drip onto the page. But even in the safety of my journal, there was one secret I could not bring myself to tell: how much I wanted my ex-husband to die. Every night I begged the fates: "Let this end. Let this be the night. Please, please, give him a massive heart attack." This death wish ate at my soul and gnawed at my pen. One morning, it poured out before I could hold it back. "There," I told the Voice, "I've said it. It's true. I do long to find him one morning with his mouth hanging open. I do." And the Voice took me gently by the hand and walked me through my dark, dark place. We talked about my secret for weeks. By the time we were finished, I had a new nightly wish—get me through this divorce—and a much lighter heart. One day, I decided that the only way to completely heal this secret was to speak it. So I told first one close friend, then another, and then, a small group. I was shocked to discover that everyone—male and female—who had been through a horrible divorce wished for the same thing! Turns out the only person who thought my secret was unmentionable was me.

If you have a secret, ask yourself, "Am I ready to break the silence?" If you're not, that's OK. Tell the Voice that you will have a nice long chat about your secret someday, but not today. In the meantime, you can still release your secret sorrows. Following the example of the Jews in Israel, you can stuff a piece of paper in your personal wailing wall. Pile a few rocks in your back yard or in the woods somewhere. Then stash a small piece of paper with a few words about your sorrow into a crack. Or burn it or bury it or drop it in the ocean. As you release the paper, simply say, "I give this to you, Spirit. Heal me." Then, don't worry about it. You'll know when it's time to break your silence, and you'll be ready.

When you are ready, surround yourself with help here on earth, too. During my darkest time, I not only wrote down my soul every morning, but I also went to private and group therapy, read healing books for an hour or more a day, and listened to spiritual masters in the car. I found a loving, supportive spiritual community that challenged me to own the power of my thinking. I did something every day to heal my soul. You don't have to plunge into the furthest reaches of your soul alone. Many have gone before you and have words of wisdom to help. If you're not sure where to get help, ask the Voice. Help will come.

Begin to Open Your Spiritual Ear

As you write your story in all its fullness, you begin to hear yourself. And as you hear yourself, you begin to discern the needs hidden behind the details of your story. And behind those needs, you begin to hear the sound of your essence, your core, your soul. This kind of listening reaches far beyond the physical act of hearing to the spiritual act of perceiving the essence within the words. Gene Knudsen Hoffman, whose work inspired the Compassionate Listening Project, explained this most elegantly in an address to the Conference of the Pax 2100 in 1994:

I am not talking about listening with the human ear. I am talking about perceiving something hidden and obscure. We must listen with our spiritual ear, the one inside, and this is very different from deciding in advance what is right and what is wrong and then seeking to promote our own agenda. We must literally suspend our belief and then listen to learn whether what we hear expands or diminishes our sense of Truth.

"Listen with our spiritual ear"—what an exquisite image. The Voice, of course, can listen only with a spiritual ear; it's the only ear the Voice has. It is we who are learning through the process of writing down our souls to find our spiritual ears and open them to the profound messages of the universe. When we do, our stories, at long last, can change.

Brian and Lisa Berman have watched people with unfathomable pain transform before their eyes. Brian described how that transformation can happen. "If you are projecting all this anger at what happened at the other person and then you go behind your words and hear what you are really protecting and defending, you can say, 'Oh, do I need to keep on protecting and defending this?' The challenge is when people are so embedded in the character in the story that they don't connect that they are the container for the whole thing. Within you is everything. Your job is to connect with that innate wisdom so you can have a deep connection with the heart of what really happened, and move on."

That's exactly what we soul writers seek: to connect with our innate wisdom, discern the truth behind our story, and move on. Move on from the place where we are to a new place where there is greater peace, greater love, greater possibility.

Early in his career, Pennebaker had a personal experience of the transformative power of telling his story, and he describes it in *Opening Up*:

My wife and I had married right out of college and, 3 years later, were questioning many of the basic assumptions of our relationship. This dark period of our life was horrible. Until that time, I had never been severely depressed. But now, on awakening every morning, the first thing I felt was an overwhelming pressure on my heart—I had to face one more day of hell.

Like many people who had never faced a major upheaval, I didn't know how to cope with a massive depression. I stopped eating, began drinking more alcohol, and began smoking. Because I was embarrassed by what I considered an emotional weakness, I avoided friends. Even though I was a graduate student in psychology, I foolishly refused to visit a therapist.

After about a month of emotional isolation, I started writing about my deepest thoughts and feelings. I remember being drawn to the typewriter each afternoon for about a week, where I would spend anywhere from 10 minutes to an hour pounding on the keys. I initially wrote about our marriage, but soon turned to my feelings about my parents, sexuality, career, and even death. Each day after writing, I felt fatigued and yet freer. By the end of the week, I noticed my depression lifting. For the first time in years—perhaps ever—I had a sense of meaning and direction. I fundamentally understood my deep love for my wife and the degree to which I needed her.

It wasn't until 8 years later that I looked back on that period in an attempt to understand why writing had been so helpful for me. Being a rather private, even inhibited person, writing helped me to let go and address a number of personal issues that I was too proud to admit to anyone. Although I hadn't talked with anyone, I had disclosed some of my deepest feelings.

Pennebaker hadn't spoken with any human person. He spoke with the all-accepting listener, what Brian Berman calls "innate wisdom," and what I call "the Voice." And just as the Compassionate Listening Project teaches, Pennebaker clearly spoke from his heart. Given the profound transition he experienced in his relationship with his wife in such a short time period, it also sounds like he found his spiritual ear and heard the inner truth of love within his external story of pain. Although no one would wish this depressive episode on Pennebaker or on anyone, it's very confirming to know that Pennebaker personally experienced the healing power of his life's work.

Find the Best Method to Access the Unconscious Mind

Pennebaker's method brings up an interesting question: do typing and handwriting provide equal access to the unconscious mind? I write both ways: by hand when I'm communicating with the Voice in my journal and on the computer when I'm writing professionally. For me, there is a big difference. Although I'm confident that I am guided and directed when I work at the computer, I have had the sensation of my hand being physically moved only when writing with a pen.

But before I stipulated handwriting over typing for soul writing, I decided to run this question past an expert on the unconscious mind. John Burton earned a doctorate in human development counseling from Vanderbilt University. He is certified as a clinical hypnotherapist and as a master practitioner of neuro-linguistic programming. He's been in private practice for twenty-three years and is the author of several books on hypnotic language. On his Web site he says, "[O]ne of the primary tasks of adulthood is to access and reclaim the connection to the infinite by learning how to use your unconscious mind." When I read this quote, I sent him an email asking if typing on the computer or writing by hand provides equal access to the unconscious. He replied:

> For me, I am not such a skilled typist, so I have to use my conscious mind to remember the placement of the keys, spelling etc. This tends to disrupt my flow of ideas from my unconscious mind to my conscious. With handwriting, I can go on auto-pilot and just let the ideas flow from my unconscious mind to the pencil and onto paper.

That sounds like a pretty clear vote for writing by hand. But many people today can type faster than Burton. So I asked him if fast typists can access the unconscious mind while staring at the computer screen. I expected him to say no, but he surprised me:

> It is my opinion that a fluent typist can remain in touch with and essentially transcribe from their unconscious mind whether looking at the screen or not. Being used to and adapted to the process can make it automatic, freeing

access to the unconscious. But, for the less fluent, we have to intermittently think about the keyboard and screen instead of thinking about our ideas. This breaks the connection requiring a re-connection. It's a matter of whatever lets your ideas flow best from your unconscious mind to your conscious mind.

For most of us, writing by hand is the best way to establish a connection with the unconscious. On paper, when a weird word or misspelling jumps out of my hand, I never slow down or cross out or edit. But on the computer, when I see a misspelled word or gibberish because my hands have slipped, I can't ignore it. Even if I don't do anything about it, I'm aware of it, and, as Burton said, my concentration is broken.

So I'm sticking with communicating with the Voice with a pen in my hand. But, if you are a fast and accurate typist, I leave it to you to experiment for yourself. Try writing to the Voice on the computer and writing by hand as a comparison. If, after a few weeks, you find you get greater access to your unconscious mind by typing, then that's the medium for you. Just remember: don't go back and edit. And don't neglect security. Store your entries in a password-protected file or delete them if necessary. Because data files are becoming another battleground in divorces and other lawsuits, you have to consider the privacy of your computer. If you are, or could be, involved in any legal tussles, look into memory-scrubbing software or write by hand and shred the pages.

Leslie found a solution to the dual problems of access to the unconscious mind and computer security by accident. After a Writing Down Your Soul class, she decided to write on the computer, but found herself struggling to stay in her unconscious mind. She was also concerned about keeping her entries private. One day, in frustration, she typed out the problem and gave it to the Voice to solve. She told me what happened next: "I was writing on the computer when it hit me to type with my eyes closed. I didn't realize it, but I had my fingers on the wrong keys. When I opened my eyes, I couldn't read a thing. It was a mistake, but it works! I know I'm talking to Spirit, and believe me, my writing is completely private. I can't even read it!"

People with arthritis have asked me how to write without using their hands. This one stumped me, so I asked Burton if you could access unconscious mind by speaking into a tape recorder. He said:

> I have tried handheld dictaphones with some success after I get past having the thing in my face. But there is the absence of access to the visual words once written. For someone with arthritis, I would suggest one of the computer programs that writes out what you speak on the screen, so you can then see what you have said.

After he sent this response, I searched speech recognition software and was floored by the demonstrations. If you have an issue affecting your hands, technology can provide other ways to write.

Honor Your Soul's Need for Expression

Many people struggle with how to surmount the emotional pain of telling their story. Gary came to a writing workshop a few months after starting Alcoholics Anonymous. Newly sober, he had plenty to discuss with the Voice, but when he tried to write, he couldn't get started. His problems seemed too vast and the pain too great to touch. One fall afternoon, while sitting in his back yard stewing over the condition of his life, he grabbed his journal and in huge black letters, screamed about a problem and swore at the Voice to fix it. With clenched muscles, Gary ripped the page from his journal. He looked around for somewhere to toss the paper, and his gaze landed on his grill. He turned it on and incinerated the page. He poured out another problem onto the page, ripped the page from the binding, and threw it on the flames. Once he started, he couldn't stop. Page after page, he demanded that the Voice hear all the horrors he was facing—everything he couldn't control, everything that was falling apart, everything that was wrong. Two hours later, he felt exhausted, but at the same time lighter, cleaner, almost holier. "I don't know if you can call that writing down your soul," he told his writing group, "but with each page going up in smoke, I felt like the Voice was taking another problem out of my life." Was he writing down his soul? We sure thought so; we gave him a standing ovation.

Gary followed his own internal direction on how to write. He transformed his journal pages into his personal wailing wall. It was wildly effective and deeply cleansing because he spoke the Voice's favorite language—the Voice's *only* language—the language of truth. He expressed exactly what he was feeling in each successive moment. He was completely present in all his pain, confusion, and fear. He screamed his story and *demanded* that the Voice listen to him. The Voice did some powerful listening that day.

Often people write for a long time when they first begin this practice, particularly if they have a tough story to tell or big problems to solve. Others tiptoe into the practice. Clark came to a writing workshop hoping writing would help him heal after his wife's death. But when he picked up his pen, words would not come. His story was just too painful. So Clark did something else. He wrote descriptions of all the rooms he and his wife were in during her last six months: doctors' offices and nurses' stations; CAT scan, MRI, and X-ray labs; operating rooms; recovery rooms; waiting rooms; hospital rooms; emergency rooms; the intensive care unit—rooms he never wanted to see again. "There were so many rooms," he told me. "I describe them for the Voice, but I do not write about my feelings. I'm afraid if I start, I'll never be able to stop." The rooms are his metaphors, his stepping-stones into the mysteries of death and loss and grief. He wrote to me after class, apologizing for not writing "correctly." I wrote back: "There is no right or wrong in talking with the Voice. You are writing in the only way that really matters—you are telling your truth. As you write about those rooms, you are doing deep rich, healing work. Keep on writing any way that brings you peace."

I've given you a lot of suggestions on how to write down your soul. But, as I told Clark, there is no *one* way or *right* way. My intent is to give you the confidence to take the plunge and begin writing yourself. In the course of writing, you will discover your own method, your own rituals, your own process. The key is simply to do it. *Show up* to have the conversation and once there, *open up* and engage deeply and fully in the moment.

You may not see the Voice on the page the first few times you write. Don't worry about that. At first, your writing is a monologue. It's supposed to be a monologue. You have a lot to say, and writing gives you a safe place to say it. In fact, nothing can happen, as the people in the Compassionate Listening Project so profoundly understand, *until* you tell your story. So tell it. But know that, although writing down your soul begins with and springs from the seed of your story, <u>telling your story is not its purpose; receiving wisdom,</u> <u>guidance, and grace is. And that requires a new kind of listening.</u>

Step Three: Listen Up

When you first begin writing, you do all the talking, and the Voice listens. But at some point your roles begin to shift. The second half of the Möbius strip illustrates the changing roles of the writer and the listener.

In step three, "Listen Up," the Voice begins to speak. The question is, of course, *how* do you, the soul writer, activate the Voice? And then, when the Voice speaks, *how* do you listen while you're writing? These are profound questions. Begin by learning to listen as the Voice listens.

Listen as the Voice Listens

Spiritual listening, like all spiritual disciplines, gets deeper and richer with practice, so don't be upset if, at first, nothing momentous seems to be happening. Just begin at the beginning and trust that you are on the road you are meant to be on and are learning exactly what you need as you need it. <u>At first, just listen to the facts</u> <u>of your experience: what happened, who did what, who said what,</u> <u>and what happened next. Listen to how you felt when it all hap-</u> <u>pened and</u> how you feel right now writing about it. Listen to your sorrows, joys, fears, and frustrations. This is the first level of spiritual listening and it is fairly easy to do. The facts and feelings come through loud and clear as you hear them in your head and then see them again a millisecond later on the page. Because of that minute delay, you get to hear your story and a tiny echo of your story. It's almost like hearing it twice. As your story reverberates in your mind, you begin to notice details and hear subtle nuances. There is great value in this first level of spiritual listening and even if you go no further, you will learn much about yourself.

But deep learning requires deep listening. To listen deeply, you must open your innermost spiritual ear. Listen as if you were a member of one of the Bermans' reconciliation workshops, listening to someone else tell his or her story. Listen to everything about the person—who in this case is you. Listen to everything about your story. Listen beyond the story to the story behind the story—the deeper joys, the deeper sorrows, the deeper fears, the deeper frustrations. Listen to the complexities and paradoxes. Listen to the little things that maybe aren't so little and the big things that, upon reflection, maybe aren't so big. Listen for how important this story was and still is and why. Listen to yourself as you have never listened before. Listen for what motivates you, what inspires you, what sets your heart on fire. Listen for what intimidates you, what frightens you, what stops you in your tracks. Listen for what keeps happening over and over again. Listen for what you want—not just what you tell your friends or your family that you want, but what you want in the recesses of your heart. When you listen with this innermost spiritual ear, you are experiencing a touch of how the Voice listens.

So how does the Voice listen? I couldn't find an answer to that question in any of the sacred texts or spiritual books in my library. But then, I stumbled upon an old friend from college—*Siddhartha*.

In this novel, written in 1922, Hermann Hesse captures perfectly both how the Voice listens and how joyful we feel when we are totally and completely heard. Substitute your Voice for "ferryman" and see if you agree.

> The ~~ferryman~~ VOICE listened very attentively. Listening, he absorbed everything, origin and childhood, all the learning, all the seeking, all joy, all woe. One of the ferryman's greatest virtues was that he knew how to listen like few other people. Without a word, the speaker felt that the ferryman took in his words, silent, open, waiting, missing none, impatient for none, neither praising nor blaming, but only listening. Siddhartha felt what happiness it is to unburden himself to such a listener, to sink his own life into his listener's heart, his own seeking, his own suffering.

What happiness indeed.

When you listen to yourself this way for ten or fifteen minutes a day for thirty days, a hundred days, a year, you cannot help but begin to hear your soul. As you listen, know that the Voice is listening, too. The Voice is always listening—and listening perfectly. It is we who are learning how to listen as we write down our souls.

||||||||||||||||||||||||||||||||||||||||||||||||||||||||||||||||||||||||||||||||||||||||||||||||||||||||

WRITE DOWN YOUR SOUL

Dear Voice,
What am I learning about listening from you?

||||||||||||||||||||||||||||||||||||||||||||||||||||||||||||||||||||||||||||||||||||||||||||||||||||||||

Be Patient

As we discovered in step two, "Open Up," life doesn't groom us for deep soul listening. So be gentle with yourself. Don't expect to understand the depth and breadth of a story the first few times you write about it. Writing down your soul isn't a one-time "aha." It's a layered discovery that reveals itself page by page over time. Don't expect or demand that your answers appear the first or the second or even the twentieth time you write. Just give yourself to the writing, and know that when you are ready you will receive the guidance and understanding you seek.

||||||||||||||||||||||||||||||||||||||||||||||||||||||||||||||||||||||||||||||||||||||||||||||||||||||||||||||||||||||||||||||

WRITE DOWN YOUR SOUL

Dear Voice,
Am I letting writing down my soul unfold,
or am I impatient? Do I have expectations
of when and how you should appear on the page?

||||||||||||||||||||||||||||||||||||||||||||||||||||||||||||||||||||||||||||||||||||||||||||||||||||||||||||||||||||||||||||||

Create Space for the Voice

For some people the Voice shows up quickly. They ask for guidance, and wham, there it is on the page. But for most, the Voice is more elusive. It isn't that the Voice isn't there. The Voice is completely present and actively listening to every word we write; it's that we don't know how to let go and let it come through. In our human conversations, we are so accustomed to dominating the conversation and pressing for what we want, that we find it hard to stop talking and be still.

Want proof? In your next conversation with your partner or child or coworker, try listening 70 percent of the time and talking only 30 percent of the time. It's tough. If you are able to pull it off, the other person will be so astonished by the dramatic change in your behavior that he or she will probably ask if you are feeling all right.

WRITE DOWN YOUR SOUL

Dear Voice,
I paid attention to how often I spoke and how often I listened in my conversations with _____. I'm a little taken aback by the results. Here's what I learned:

Here's an even more revealing listen/talk experiment: Observe yourself the next time you pray. Pay attention to how much of the time you are busy stating your case, listing your woes, justifying your behavior, or begging for something and how much of the time you are sitting quietly in the presence, feeling grateful, and just being open. The truth is, we yak our way through prayer, just as we yak our way through our human relationships, telling the divine not only what we want but how it should be delivered. And so it is with deep soul writing. We are so accustomed to doing the talking and moving the conversation toward a particular desire or outcome, that we find it difficult to create room for the Voice to step in.

WRITE DOWN YOUR SOUL

Dear Voice,
This week I paid attention to how often I did the
talking when I prayed and how often I was still.
How much room do I give you to communicate with me?

||||||||||||||||||||||||||||||||||||||||||||||||||||||||||||||||||||||||||||||||||||||||||||||||||||||||||||||||||||||||||||

Consciously Want to Hear the Voice

There are several ways to create space for the Voice. The first is to consciously want to hear it. This is not a small thing, nor is it a given. Reading this book doesn't automatically mean you want to see the Voice on the page—because when you give the Voice room, the Voice *will* speak, and you will hear things that you may not expect or want.

One morning in 1997, I was screaming in my journal about all the terrible things my husband was doing: "We are at war! War over visitation, war over clothes, war over little league, war over the piano, war over birthday presents, war over vacation, war over summer camp, war over money, war over dishes, war over toasters, war over *nothing*, but war nonetheless!"

After pages of venting, I *demanded* that Spirit do something about my ex: "Give me a little advice. I never wanted an enemy, but I have one. What do I do about my enemy? How do I defeat him? How do I get him to lay down his arms?"

During this rough time, I'd been devouring the Psalms and reveling in what a great job Yahweh did smiting and smoting the enemies of his people. I figured I was one of God's people, too, so surely God would handle my enemy for me. I know this sounds insane, but when you're consumed with anger—and haven't yet done enough soul writing to understand that your anger is really about you—you'll ask the Voice to do some outrageous things. In the millisecond before the next words came out on the page, I felt this little twitter of joy because I truly believed that I was about to get some really valuable, biblical-quality advice on how to win. Then, my hand wrote: "Lay down mine? Stop fighting? Love my enemy?"

I was appalled. This is *not* what I wanted to hear. And I said so. "Oh, dear God, anything but that." (When I tell you I've wrestled with the Voice, I'm not kidding.) Just in case the Voice had somehow missed the severity of our situation, I wrote another laundry list of even more rotten things my husband was doing, and, thinking that surely this time I'd stated my case in a totally compelling and convincing way, I asked again. "How do I get him to lay down his arms?"

Instantly on the page, I got my answer: "Lay down mine? Stop fighting? Love my enemy?"

The Voice, of course, was right, but three years passed before I was willing to follow that guidance. I don't tell you this story to scare you away from writing down your soul. I tell you this story to illustrate dramatically that when you consciously want the Voice and create space for the Voice to respond, the Voice not only comes through, it comes through unmistakably. "Love your enemy" most certainly did not come from me—at least not from my conscious self. It came from a much deeper place in my soul where Truth with a capital *T* resides.

So *how* do you consciously want the Voice? It isn't something you do. It's something you intend. It's something you allow. It's something you welcome. And it's something you surrender to. The Voice extends everyone an open, standing invitation to connect. Your conscious desire to hear and see the Voice on the page is your acceptance of that invitation.

|||||||||||||||||||||||||||||||||||||||||||||||||||||||||||||||||||||||||||||||||||||||||||||||||||||||||||||||||||||||||||||

WRITE DOWN YOUR SOUL

Dear Voice
Am I ready to accept your invitation? Do I really want
you to talk with me? How do I know I'm ready?
What do I need to do to truly open to your Voice?

|||||||||||||||||||||||||||||||||||||||||||||||||||||||||||||||||||||||||||||||||||||||||||||||||||||||||||||||||||||||||||||

Ask for an Understanding Heart

If consciously wanting to hear the Voice is your RSVP to its invitation, asking for understanding is comparable to walking in the door and seeing the feast the Voice has prepared for you. Just imagine how different my experience with the Voice would have been in 1997 if I had asked for understanding instead of victory.

Solomon, the Old Testament king, understood the importance of understanding. His father, the Jewish king David, had named Solomon his heir to the throne just before he died. Solomon then married the daughter of the Egyptian pharaoh and moved to Jerusalem to begin construction of a temple and his palace. At that point life looked pretty spectacular. But Solomon understood that he wasn't equipped to be king. He turned to Yahweh and prayed:

Now, Yahweh my God, you have made your servant king. . . . But I am a very young man, unskilled in leadership . . . . So give your servant an understanding heart, to govern your people, to discern between good and evil, for how could one otherwise govern such a great people as yours? (I Kings 3:7-9)

What a simple, beautiful, and profound prayer: give me an understanding heart. It was, by the way, the right request. Not only did Yahweh give Solomon the wisest heart ever known, Yahweh gave him things he *hadn't* asked for, like riches, glory, and a long life.

It is the understanding heart that does the deep listening. Your understanding heart is one and the same with your innermost spiritual ear. When you ask the Voice for understanding, you create a space in your heart for fresh awareness, new learning, real discernment, and profound seeking. When you ask for understanding, you express a willingness to go deep to tap into the Truth of the universe.

||||||||||||||||||||||||||||||||||||||||||||||||||||||||||||||||||||||||||||||||||||||||||||||||||||||||||||||

WRITE DOWN YOUR SOUL

Dear Voice,
What does it mean to have an understanding heart?
How would my life change if I had an understanding heart? How can I ready my heart for understanding?

||||||||||||||||||||||||||||||||||||||||||||||||||||||||||||||||||||||||||||||||||||||||||||||||||||||||||||||

Trust That You Are Safe and Loved

Asking for an understanding heart is not something you do lightly. Because when you create a space for the Voice, you become vulnerable in a way you have never allowed. Even in our intimate relationships, there is a part of us held in reserve. But in relationship with the Voice, there are no secret corners, no feelings kept hidden, no desires left unspoken. Even if you think you're holding back, all is known, all is seen. You are fully and completely naked.

This nakedness would be unbearable without faith. Not faith in the religious sense; connection with the Voice is not tied to any belief system. This kind of faith is a deep knowing, a deep trust that all is intended for the good—a trust that you are safe speaking your truth, safe asking for understanding, safe creating space for the Voice to speak and safe receiving the Voice's guidance. It is a deep trust that you are protected and guided, and safe and loved. If you have to choose just one thing to trust, this is it. Say, "I am safe and loved." And know that this is true.

|||||||||||||||||||||||||||||||||||||||||||||||||||||||||||||||||||||||||||||||||||||||||||||||||||||||||||||||||

WRITE DOWN YOUR SOUL

Dear Voice,
Do I trust that I am safe and loved? Do I?

|||||||||||||||||||||||||||||||||||||||||||||||||||||||||||||||||||||||||||||||||||||||||||||||||||||||||||||||||

Be Present in the Now

Lisa Berman said that spiritual listening requires being 100 percent present in the now because "that which wants to be heard is now here." The events of a person's story may have happened a long time ago, but the pain the person feels about it is still happening right now. The reflective bowl of listeners that holds the story is also here right now, and the awareness of the story behind the story happens in the moment. So the healing happens the only place it can happen—right here and right now.

This is also true for writing down your soul. You might write about what happened when you were ten or what happened yesterday. You might write about falling off your bike at five or arguing with your spouse last night. Or you might write about the future—what could or might or should happen. You can be all over the calendar in the content of your writing, but you can hear the Voice only in the present, today, this very moment. You can learn only in the present. You can receive guidance only in the present. You can change your mind only in the present. You can heal only in the present.

What happened, happened. All the writing in the world won't and can't change the outcome of the story: She left me. I lost my job. My mother died. My husband was sued. I got sick. My father closed his business. My boss lied. The house didn't sell. Someone else got the promotion. The doctor was wrong. The market tanked. My daughter dropped out. Writing can't change the story, but it can change how you think about it. It can change how you feel about it. It can change what it means to you. It can expose little nuggets of truth buried in the sand of your story. It can change the spiritual lessons learned and the spiritual risks taken. It can change your relationship with the people involved, and it most certainly can change your relationship with Spirit. But all that change can happen only today, right now, in the present moment.

If the importance of being in the now is unclear to you, consider the nature of forgiveness. Every spiritual tradition teaches the power and necessity of forgiveness. Whatever it is that needs to be forgiven already happened. But forgiveness can happen only today. You can't go back and pretend that you forgave someone ten years ago. And you can't project yourself forward and decide to forgive someone three years from now. You can only forgive right here, right now, this moment.

It's obvious that forgiveness can only happen in the now, but "nowness" also applies to everything. Everything happens in the now because that's where you are. You really can't be anywhere else. You can visit another time and place in memory of the past or projection of the future, but you can live only in the present. Rabia of Basra, the popular Islamic saint, understood this idea way back in the 700s: "The soul does not understand the word *seasons*. The petals on the sun can only be touched by now." I love the short, blunt way Rumi, the great mystic Sufi poet, says it: "Why lay yourself on the torturer's rack of the past and future?"

The present is also where the Voice is. The Voice can't get stuck in the past or get lost floating around in the future. The Voice is always present in the perfect, infinite, transcendent now.

You access the Voice by stepping beyond your everyday conscious life and entering the portal of the unconscious mind. The unconscious mind, it's important to know, has no awareness of past or future. For the unconscious there is only now. For the unconscious mind, the pain didn't happen last week or last year or last decade—it's happening right now. For the unconscious mind, a scary thing isn't happening in the future; the fear and worry about it is happening right here and right now.

So if you want to meet the Voice, be open and fully present in the now, and "that which wants to be heard" will make itself known.

|||||||||||||||||||||||||||||||||||||||||||||||||||||||||||||||||||||||||||||||||||||||||||||||||||||||||||||||||||||||||||||||||||||||||||||

WRITE DOWN YOUR SOUL

Dear Voice,
Where is my focus?
Is it in the present or the past or the future?

|||||||||||||||||||||||||||||||||||||||||||||||||||||||||||||||||||||||||||||||||||||||||||||||||||||||||||||||||||||||||||||||||||||||||||||

Be Willing to Explore Beyond Conscious Mind

If "that which wants to be heard" was already in your conscious mind, you already would have "heard" it. You already would know it. It wouldn't have to do anything to get your attention or make itself known. So this information that "wants to be heard" must be new and it must be coming from somewhere beyond your conscious mind. Which leads to a momentous question: What's beyond conscious mind? Well, religion and science have been wrestling with that for eons. Here's a brief introduction to some of the possibilities.

Let's start with your personal subconscious mind. The subconscious mind is vast—more vast than we can comprehend. In his bestselling book, *The Biology of Belief*, Bruce Lipton tries to describe it: "When it comes to sheer neurological processing abilities, the subconscious mind is millions of times more powerful than the conscious mind." He puts two pictures side by side to illustrate the information-processing powers of the subconscious and conscious minds. The first picture, a complex, 20-million-pixel photograph of Machu Picchu, represents the amount of information the subconscious mind absorbs in one second. Next to it is a black rectangle of the same size with one barely visible, 40-pixel white dot in the center, representing the amount of information that enters the conscious mind in that same second. Just try to imagine the amount of information accumulated in your subconscious mind in all the seconds in your life thus far!

Beyond your personal subconscious mind is the collective unconscious filled with all the symbols and archetypes accumulated by humanity. This is where the symbols in your dreams come from. And beyond the collective unconscious is an even greater storehouse that holds the knowledge of all that is or was or ever will be. We humans clearly believe this massive storehouse exists because, regardless of where we live on the planet or in which century, we create names and images for it.

Trees and books were the most common symbols in ancient cultures for this cosmic storehouse. The Hebrews called it the Book of Life. The Book of Life appears for the first time in the Torah when Moses admits that his people have been worshiping other gods. He doesn't want Yahweh to punish them, so he goes for an all or nothing negotiation: "[I]f it pleased you to forgive their sin. If not, please blot me out of the book you have written" (Exodus 32:32). In Psalm 139, King David gives an eloquent and amazingly detailed description of the information in the Book:

Yahweh, you examine me and know me,
you know when I sit, when I rise,
you understand my thoughts from afar.
A word is not yet on my tongue
before you know all about it.

You created my inmost self,
knit me together in my mother's womb.
Your eyes could see my embryo.
In your book all my days were inscribed,
Every one that was fixed is there.

The Qur'an speaks of a similar book:

No one can die except by God's permission, according to
the Book that fixeth the term of life. (003:145)
No woman bears or is delivered except by His knowledge,
nor does he who is aged reach old age or is aught dimin-
ished from his life, without it is in The Book. (034:011)

Given that Exodus was written around the sixth century BCE and
the Qur'an around 650 CE, this archetypal symbol of the Book of
Life spans over a millennium.

The symbol of a tree is even older. In the second Judaic creation story, written in approximately 950 BCE and recorded in
Genesis 2:9, all information is stored in the Tree of Life and the
Tree of Knowledge. In the Buddhist tradition, Buddha found enlightenment sitting under the Bodhi tree 2,500 years ago, and from
then on, that tree symbolized pure transcendent knowledge.

In the Hindu tradition, the seat of all knowledge, Paramatman, is not a book or a tree; it is the absolute truth that dwells
somewhere deep within. In ageless Inuit mythology, it is Silla, the
formless ether that is the primary component of everything and
everyone. It seems that every spiritual tradition, going back to in-
digenous oral cultures, has attempted to describe this place of all
knowledge.

Poets have tried a hand at naming it, too. William Butler Yeats called it "Spiritus Mundi" in his famous poem "Second Coming." Ralph Waldo Emerson called it the "over-soul." In 1841, he wrote, "Within man is the soul of the whole; the wise silence; the universal beauty, to which every part and particle is equally related, the eternal *one.*"

The mystics, of course, have always talked about this eternal wholeness or oneness. In the fourteenth century, Meister Eckhart, the renowned German monk and scholar, reported this profound conversation with an ant:

> Having lunch in a field one day, I troubled an ant with some questions. I asked of him humbly, "Have you ever been to Paris?" And he replied, "No, but I wouldn't mind going." And then asked me if I had ever been to a famous ant city. And I regretted that I hadn't, and was quick to add, "I wouldn't mind, *too!*" This led to a conclusion: There is life that we do not know of. How aware are we of all consciousness in this universe? What percent of space is this earth in the infinite realm? What percent of time is one second in eternity?

The poet Rumi called this cosmic place of true knowledge a field:

> Out beyond ideas of wrongdoing and rightdoing,
> there is a field. I'll meet you there.
> When the soul lies down in that grass,
> the world is too full to talk about.
> Ideas, language, even the phrase "each other"
> doesn't make any sense.

For centuries this ethereal storehouse of knowledge has been relegated to the realm of religion. It was well and good for people to *believe* the Book of Life or Tree of Knowledge or Paramatman existed, but no one could *know* that they existed. These things were a matter of faith, not science. But when Albert Einstein postulated that energy and matter were one and the same—that is, that the raw material of the universe is *non*material—he set off a revolution in scientific thinking that continues to draw the spiritual and scientific realms closer and closer together. Science, it appears, is discovering what the ancients, mystics, and poets always knew.

The Hungarian systems theorist Ervin Laszlo is at the forefront of scientists exploring this integration of the scientific and spiritual. Laszlo's research carries Einstein's theories forward into an exploration of the quantum vacuum or zero-point energy. Zero-point energy is the residual energy that remains when all other energy has been removed. This remaining field of energy is the invisible web running in the background of the universe, connecting everything to everything else. In the zero-point field, all is indeed—as Ralph Waldo Emerson speculated—*one*.

Laszlo explained the implications of the zero-point field in "The New Scientific Paradigm," a speech he gave in 1996:

> Matter as well as mind evolved out of a common cosmic womb: the energy-field of the quantum vacuum. The interaction of our mind and consciousness with the quantum vacuum links us with other minds around us. . . . It "opens" our mind to society, nature, and the universe. This openness has been known to mystics and sensitives, prophets and meta-physicians through the ages. But it has been denied by

modern scientists and by those who took modern science to be the only way of comprehending reality.

Now, however, the recognition of openness is returning to the natural sciences. Traffic between our consciousness and the rest of the world may be constant and flowing in both directions. Everything that goes on in our mind could leave its wave-traces in the quantum vacuum, and everything could be received by those who know how to "tune in" to the subtle patterns that propagate there . . . it is as if something like an antenna were picking up signals from a transmitter that contains the experience of the entire human race.

Those "wave-traces in the quantum vacuum" sound exactly like the subtle substance on which the Akashic Record is traced. The Akashic Record is a modern term for the universal filing system that contains every feeling, thought, word, and action of every soul. The records are described as being imprinted on akasha, a Sanskrit word that means primordial ether or cosmic substance. In our minds, we may picture the Akashic Record as a giant library or computer, but of course it isn't anything physical, just as the Book of Life and Tree of Knowledge were never limited to a physical book or tree.

Lauralyn Bunn trains people to tune in to and retrieve signals from the Akashic Record which, she says, "is an all-encompassing place of knowledge that is loving and supportive but also neutral, unbiased. It has no impetus to influence or change our minds. It provides us with information about the present and the past, but it leaves us at the place of making our own decisions. The law of free will is never violated."

I first became aware of the Akashic Record and Lauralyn Bunn through Dr. Hardt and the Biocybernaut Institute where Robert and Michelle Colt went for brain-wave training. When one of Lauralyn's students told Hardt that she began the process of accessing the Akashic Record by saying a sacred prayer, Hardt asked her to say the prayer. As she did, the brain-scan equipment captured a huge shift in her brain-wave pattern. Hardt immediately recognized the unique, mystical theta-wave pattern because he'd seen it repeatedly in the brain scans of Zen masters. In his book, *The Art of Smart Thinking*, Hardt describes what mystical theta can do:

> [W]hen mystical theta waves are present in our brains, we
> are in touch with that inner problem solver that resides in
> all of us, the subconscious, intuitive mind. This part of us is
> somehow able to access dimensions of reality that our con-
> scious mind cannot . . . in the mystical theta state, the mind
> is able to access a universal filing system that records every
> thought, word, and action in the collective human memory
> hard drive, a universal data base in the quantum field.

I wanted to hear what my "inner problem solver" had to say, so I asked Bunn for an Akashic Record reading. With her help, I posed questions to the masters and teachers who have access to my soul's experiences in the Akashic Record. The masters and teachers re-trieved the information and conveyed it back to Bunn symbolically. She then translated the meaning of those symbols as closely as she can into English.

I started the reading by asking, "I see now that my ex-husband's behavior forced me to discover deep soul writing. Was that our purpose together?"

Their response was swift and emphatic: "Nothing was forced. This was very much a free will choice."

Lauralyn could sense my surprise and she explained that the masters and teachers were responding to the living energy of the word *forced*. She also stressed that the Record holds my soul's information but nothing is predetermined. There is always free will choice.

One of the questions I asked the masters and teachers was, "When I or any other deep soul writer writes, where are we? Are we in the Record?" They said that we are not in one place; we can be in many different places. We are not always in the Record, but we *can* be in the Record. To help me understand, they gave this analogy:

When you are standing in a meadow of wildflowers in the Alps or standing on the rim of the Grand Canyon, you feel deeply connected. Are you at that moment only in the Alps or only in the Grand Canyon? In the same way with writing, in a moment of deep connection, there is a sense of awe. There is no question that you are connected. If someone said, "Prove to me that you are connected with something," you'd say, "I don't have to. I know." It is only the human mind that prevents the deep connection all the time. The potential to be connected is always there.

This analogy was so beautiful and so perfect that I asked if I could use it in this book. They said they'd be honored. Use it, they said, because it's important to demystify the mystical experience of writing down your soul. "Make it like brushing your teeth," they said. (Don't you love that? Deep soul writing is a daily practice that keeps you spiritually healthy—like brushing your spiritual teeth!)

I asked several more questions that made it clear that they were reading *my* soul's record. As the session ended, I asked one last question about my mother, who had died a few months before, "Have I uncovered all the gifts she had for me?" There was a long pause, followed by a faint "You already know that." The masters and teachers were right. I had spent months exploring my relationship with my mother in my deep soul journal. Lauralyn explained that their response was in direct proportion to the strength of my need to know.

From my Akashic Record reading, I learned that there *is* a Book of Life. It is real and it is accessible. I learned that words carry distinct energetic vibrations. I learned that the universe responds in equal measure to the strength of your desire to know. And I learned that the information available to you as you write down your soul is endless.

Ask Questions

Out of the vast range of possibilities from your personal subconscious to the collective unconscious to the Akashic Record to the quantum field, how can you retrieve the specific information you need to understand and improve your life at this moment in time? Out of all of the information available, how can "that which wants to be heard" make itself known?

One profound way: Ask questions—lots and lots of questions. Just as wanting to hear the Voice is comparable to accepting the Voice's invitation, and asking for an understanding heart is comparable to getting a peak of the vast array of dishes available, asking questions is analogous to sitting down and beginning to sample the feast that is your soul.

Questions are the Mars Explorers of your psyche, flying out from your conscious mind to probe the vast vaults of information available in your subconscious mind and beyond. They are magnets attracting information that fits, that makes sense, that just might be that thing we call an answer. But the relationship of questions and answers isn't that simple. In the spiritual practice of writing down your soul, there is no one-to-one relationship between a question and its answer.

If there were one perfect question, we'd all be scrambling to discover it, ask it, get the perfect answer, and be done with our soul's exploration. Obviously, that's not the way life or questions work. One question just seems to lead to another and another and another. We can accept that there are thousands of questions. But it isn't so easy to accept that for each of those questions, there is no one "right" answer. Many of us spend decades chasing those elusive "right" answers, and the whole time life just seems to get more and more complex.

This complexity is a gift. Not knowing the right answers is a gift. Problems without apparent and immediate solutions are gifts. It is these gifts that bring us to the place where we want and need a spiritual solution. It is these gifts that cause us to cry out for help. It is these gifts that bring us to the page and to the Voice. And it is these gifts that provoke us to start asking new questions— questions that don't have one right answer.

When you find yourself blurting out questions you've never asked before, questions that provoke you and possibly even scare you, be thankful. You are exiting your conscious mind and probing the unconscious. As the questions get more and more profound, your probe is going deeper and deeper into the many layers of information that are unseen but always available. Ask important questions, and answers—varied, rich, potent answers—are on the way.

Jesus said it most elegantly and succinctly:

> Ask, and it will be given to you; seek, and you will find; knock, and the door will be opened to you. For everyone who asks receives; everyone who seeks finds; everyone who knocks will have the door opened. (Luke 11:9–10)

Asking by its very nature means posing questions. But there are questions that leave you stuck and questions that open your soul. There are questions that move you an inch and questions that propel you a mile. How do we formulate questions that will help us the most?

Avoid Unproductive Question Formats

When I spoke with Brian and Lisa Berman about questions, they stressed that in Compassionate Listening workshops participants are instructed to ask one another open-ended questions and avoid questions that can be answered with a simple yes or no. This wisdom applies to writing down your soul as well. Asking the Voice a yes-no question only produces frustration. Even if you should see a yes or no answer on the page, you won't know for certain if you should take it or how to implement it, because there's no supporting dialogue. As a rule, the Voice is not a big fan of short, blunt, do-this or don't-do-that answers. The Voice is interested in expanding the depth and breadth of your heart and soul.

When you find a yes/no question bubbling onto the page, don't ask the Voice; ask yourself. You'll find that answering provocative questions like, "Am I ready to hear you?" or "Do I really want an understanding heart?" in the presence of the Voice propels you to dig deep and find the truth.

"Why?" is another tough question to stop asking because it's the one thing we all want to know. Why didn't I get the job? Why did the cancer come back? Why am I broke? Why did he do that to me? Why did she leave me? Why is this happening to me? All of these questions can be distilled down to "Why me?" And "why me" isn't very helpful. It's really a form of whining that seeks to know whom to blame. The truth is, there is no one to blame. Things are not happening *to* you but *for* you. In the midst of chaos and pain this distinction can feel like an electric shock. You may feel like screaming, "Whadya mean, this is *for* me?!" But the truth is, your soul has called this forth. The sooner you give up the victim mentality of "why me," the sooner you can dive into and explore your soul's truth. If you want to ask "why me," substitute "Why did my soul call this forth?" When you do, be prepared for a jolt. What appears on the page next may shock you.

Another type of question that doesn't work is asking about the future. When will I find my true love? When will they sign the contract? When will he stop drinking? Will I win the lottery? When people ask these kinds of questions, Lauralyn Bunn tells them that "time is the least solid type of information that comes from the [Akashic] Record because time is tied to shifts in consciousness." In other words, whatever you're asking about will happen when you are ready. So if you want to ask "When will I find true love?" change it to "What needs to happen for me to be able to attract and meet a loving partner?" If you want to know "When will he stop drinking?" change it to "How did I get in this situation?" Whenever a question about the future creeps into your writing, ask yourself, "What shift in consciousness is this really all about?" and ask how you can make that shift.

A fourth stream of questions that doesn't work well is asking about the other person. Writing down your soul is about you—your understanding, your spiritual development, your soul's unfoldment—not the other guys. The focus is on what you are learning, what you are experiencing, what you are discovering about yourself. The person being healed, expanded, nurtured, and loved is you—not the other guy. That doesn't mean that the other person can't or won't change. As your soul shifts and changes in the course of writing down your soul, another person may well shift in a dancelike response, but don't write with the intention of altering another person's behavior. Do it with the intention of deepening your understanding of yourself. Do it with the intention of exploring your own soul. Do it with the intention of having a personal conversation with the Voice. Don't ask what's wrong with the other guy, ask about yourself.

ımmıllılımıliylymyllılımyımıımımıımıllılymyllıymyımıllıımymmıllılımyy

WRITE DOWN YOUR SOUL,

Dear Voice,

Truth be told, I wish I could have yes/no answe
And I do want to know "why me?" And I do wa
look into the future, especially if it s a good
and I really, really do want to know what is
with ___. But I'm willing to let go. I'm
to ask different questions. Tell me
kinds of questions will help me the

ımmımımımıımıımıllılıymymıımımımımımımıllıymıımıllıımyy

Use Powerful Question Formats

In a Compassionate Listening workshop, members are taught to ask "compassionate questions." In a compassionate question, the focus isn't on finding the facts, drawing conclusions, or making judgments; the focus is on connection—connection with the truth of your story, connection with the truth of your soul, connection with the truth of others, and connection with the divine within. Compassionate questions move you closer and closer to your whole and holy self.

When I heard Brian and Lisa talk about "compassionate questions," I got out a list I'd been building of over 200 soul-sparking questions that I'd heard or read or written myself. As I stared at the questions, I noticed that they fell pretty naturally into five related groups with five distinct purposes: Becoming Aware, Understanding and Meaning, Soul Exploration, Imagining and Incubating, and Creating and Manifesting.

As you read the sample questions in each category, notice how each question acts like a strange attractor. As you ask it, you can practically feel it gathering force as it energizes and attracts information from your conscious, your subconscious mind, the collective unconscious mind, and beyond. Each question is a bridge between where you are and where you want to be, what you know and what you want to know, who you are and who you want to become. Notice, too, that these questions can be repeated a day, a month, a year later. These are soul probes that penetrate further and further into your core each time you ask them. These are not questions you ask once and mark "done."

Take your time with these questions. Any one of them can consume several days of writing time. Don't rush. This isn't a race to spiritual enlightenment; it's an exploration of your soul. Choose a question that speaks to you and work with it until you feel ready to move on. Chew over what comes out. And ask lots of follow-up questions.

Don't ask a question because you think you should. Don't ask a soft question because you're afraid to see the answer to a hard one. Don't sidestep. Ask what you really *really* want to know. Ask for your soul's truth. If that means putting a gutsy, tough question out there, go for it. The big Voice will respond in equal measure to your small voice. And the power of its answer will be in direct proportion to the energy of your question.

Don't worry about an end result. Release the need to find the "right" answer. If you write with an open ear, an understanding heart, and a deep knowing that no matter what you write, you are safe and loved, the information your soul seeks will make itself known.

Initially, you may start the dialogue with some of these sample questions, but soon you will find yourself following where the Voice leads. One question will lead to another and another, and the conversation will flow. Information you never knew was inside you will appear on the page. You will not have to strain to hear "that which wants to be heard."

Category 1: Soul Questions that Support BECOMING AWARE: Becoming aware is the ideal place to begin your conversation with the Voice. These questions help you become conscious of what's happening in your life and how you feel about it. Don't assume you already know this information. You don't. Not really. Not deeply. It's easy to spit out a few facts and state your feelings, but that's just the story and feelings on the surface. In communion with the Voice, you can explore the story behind the story, the feelings behind the feelings, the truth behind the facts. You may think at first that these are baby questions and you can skip them; do not be deceived. They are simple, but they are profound. They are your soul's essential first steps. They help you get grounded in your story—your full story.

STUCK

Where do I feel stuck? What isn't moving?

Is this block in me or out there?

If it's in me, how did it get there? If it's out there, how is it affecting me?

Why is being stuck bothering me so much?

What do I think will happen when I'm unstuck?

IN FRONT OF ME

What's in front of me right now? What do I need to deal with or face?

Is there a decision that needs to be made?

How do I know that *that's* the decision that needs to be made?

What do I need to know, do, be, or have in order to make a wise decision?

How can I get what I need?

FEELING

What am I feeling—really feeling—right now?

What feeling lies behind that feeling? And behind that feeling?

What are my feelings telling me?

SEE

What have I been unwilling to see?

What have I been hoping would go away?

What has to happen for me to start seeing more clearly and fully?

SKIN

How comfortable am I in my skin, being who I am?

When am I comfortable? When am I not?

What makes the difference?

What has to happen for me to become more comfortable with myself?

PARTS OF ME

What parts of me are joyful, happy, content?

What parts of me are angry, sad, depressed?

What parts of me are trapped, frustrated, afraid?

What parts of me do I like? Not like? Loathe? Love?

What parts do I want more of? Less of?

HAPPY

On a scale from one to ten (ten being high) how happy am I right now? Why did I pick that number?

What is that number telling me?

What elements in my life generate happiness, and what elements do not?

What is the happiest moment of my day? What does it mean that that particular moment makes me happy?

What is my happiest relationship? Happiest goal? Happiest activity?

STRESS

What are my stressors (all of them)? What are their names?

On a scale of one to ten (ten being high) how intense is each one?

How does each one manifest itself? When does it hit me? Where do I feel it?

How do I know I'm under stress?

How is stress impacting me? My life? My relationships? My desires?

Is my stress getting better or worse?

What has to happen for me to reduce stress or handle it more effectively?

RIGHT NOW

Deep down, why am I talking with you, the Voice, right now?

What do I need to learn, discover, uncover right now?

Why is writing down my soul an important spiritual practice for me right now?

CHANGE

What change is headed my way like a train coming down the tracks?

How do I know it's coming?

Is this the first time I've recognized it? Acknowledged it? Admitted it?

How do I feel about this change?

How can I prepare myself?

WORRY

What am I worrying about?

When do I worry? What do I do when I'm worrying? How do I know I'm worrying?

What has my worrying produced? What has changed?

What do I get out of worrying? Why do I persist?

I must be getting something out of worrying. What is it?

What has to happen for me to stop worrying?

||||||||||||||||||||||||||||||||||||||||||||||||||||||||||||||||||||||||||||||||||||||||||||||||||||

WRITE DOWN YOUR SOUL

Dear Voice,

Reading these questions, I realize I'm not as aware of myself as I want to be. Help me. How do I begin to get more in touch with what's going on inside?

||||||||||||||||||||||||||||||||||||||||||||||||||||||||||||||||||||||||||||||||||||||||||||||||||||

Category 2: Soul Questions That Support UNDERSTANDING AND MEANING: These questions take your story to the next level—the meaning within the story. They support you as you seek to understand what is happening and what it all means. They are right there with you as you connect the dots between what's happening in your life and how you attract, create, or allow it. These questions are not easy, and they typically don't feel good. But don't turn your face away from them. As you ask these kinds of questions you are building those new neural pathways the Colts spoke about. You are literally changing your mind. Want to improve your life? Start here.

TURN IN THE ROAD

Where am I?

How did I get here? What steps led me to this place?

At what point did I take a turn in the road? What was that turn? Why did I take it?

What options existed? Why did I reject or refuse to see the other options?

DISTRACTIONS

How have I prevented myself from looking at the truth?

What have I done to distract myself?

While I was looking away, what happened?

STRENGTH

What will I see if I open my eyes and really look completely and honestly at my situation?

What kind of strength will it take to look that honestly?

Do I have that kind of strength?

What can I do to become strong enough to take a long hard look at where I am?

CREATE

How did I create this situation?

What was I thinking?

How did my thoughts, my beliefs, my actions produce this situation?

TOXIC

What thoughts of mine are toxic?

What damage have those toxic thoughts done?

I need new transformed, healthy thoughts. What are they?

What toxic words and language do I repeatedly say?

What words do I wish I could unsay?

What damage have those words done?

I need new words. What are they? What do they sound like?

What toxic things have I done? What do I wish I could undo?

What damage was done? What would I do differently today?

Why do these toxic things keep repeating?

Am I ready and willing to change?

What do I have to do to transform my thoughts, words, and actions?

BLOCKS

What is preventing me from expressing my gifts, finding my way, living the life I say I want? What is blocking me from speaking my truth and acting on it?

What has to happen for me to step over the blocks, push them aside, or transform them?

Why have I not been willing to do that work before? If I am now, what changed? If I'm still not willing to do that work, what's that all about?

What needs to happen for me to become willing?

PATTERNS

In what ways are the things that happen to me related?

What is the recurring pattern? The theme? What is its name?

How does this theme keep showing up?

I must be getting something out of this pattern, or I would not keep repeating it. What am I getting out of dealing with the same problem over and over?

Do I want to stop? Really?

What has to happen in order for me to stop having this same experience over and over?

TRIGGERS

What triggers fear in me? Anger? Frustration? Depression?

Is it something in my environment or relationships, or is it inside of me?

Where does it come from? When did I first experience this trigger?

How can I recognize the trigger? Where does it start in my body?

How do I react to it?

How can I become conscious of this trigger so it doesn't have dominion over me?

When it comes again, how can I recognize it and what can I do differently?

TELL MYSELF

What do I tell myself about myself?

How does it make me feel? Is it true?

Do you, the Voice, think it's true?

What do I want you, the Voice, to say to me about me?

If I put it in a sentence could I say it to myself out loud every day? Could I? Would I? Will I? When?

What do I do because I have to?

Why do I feel I have to do all these things?

What makes these things "necessary"?

What do I choose to do?

Is there a difference between what I choose to do and have to do? What's the difference?

What would happen if I did the things I "choose to" instead of the things I have to?

What has to happen for me to start "choosing to" instead of "having to"?

TRUE/FALSE

What is false for me in my story, my choices, my interpretation of what happened?

What is true?

What's the difference? How can I tell the difference?

||||||||||||||||||||||||||||||||||||||||||||||||||||||||||||||||||||||||||||||||||||||||||||||||||||||||||||||||||||||||||||||||||||||||||||||||

WRITE DOWN YOUR SOUL

Dear Voice,

This is a profound group of questions for me. I do want more understanding. I do want to get more meaning out of what happens in my life. I know my life can't really change until I do. Help me understand what these questions are telling me. Tell me, what do I need to be asking myself right now?

||||||||||||||||||||||||||||||||||||||||||||||||||||||||||||||||||||||||||||||||||||||||||||||||||||||||||||||||||||||||||||||||||||||||||||||||

Category 3: Soul Questions That Support SOUL EXPLORATION: These questions probe the inner reaches of your soul. They are like deep-sea robots that explore trenches on the ocean floor—only the ocean in this case is you, your core, your soul. Have you seen photographs of the deep-sea life forms those submarines have discovered? Some of the new animals have bizarre shapes, scary fangs, and eerie body parts that generate their own light in the blackness. Your soul exploration may unearth some unexpected shadows, deep fears, and puzzling self-defeating patterns. But keep going. You will discover that the light and love of the universe glows deep within your soul.

ORPHAN

What parts of me have I been unwilling to acknowledge?

Am I willing to acknowledge them now? Embrace them? Welcome them into my heart? Why or why not?

What would happen if I decided to love and embrace all of me?

WAITING ROOM

When something upsetting happens, do I stay present and aware and deal with it, or do I retreat into a mental "waiting room" until things calm down?

What kinds of things send me running to my waiting room?

How much of my life do I spend in this waiting room?

How well is this waiting room strategy working for me?

What can I do to become aware and fully present instead of hiding and avoiding?

What needs to happen for me to stop hiding in my waiting room?

BREAKING A PATTERN

I think there's a pattern in my life, and I don't want to
   perpetuate it. What is the pattern? Where does it keep
   appearing?

When did it start? How has it evolved?

In what ways am I passing it on to the people around me?

Why do I want to break it—or why not? What price am I
   willing to pay?

What needs to happen for me to end this pattern?

MASKS

What mask do I present to the world? Myself? My family?

How many masks do I wear?

Do people see only the mask, or do they see through it
   to me?

When am I me? Am I ever maskless?

When did I start wearing masks? Why did I start?

Which mask do I want to take off first? Second? Third?

How painful will it be to take off these masks?

How do I take a mask off?

How do I prevent myself from putting a mask back on the
   first time things get rough?

WHO I AM

Who am I when I am fully, completely, divinely me?

Who did Spirit create me to be?

✓ AFRAID

What am I afraid of?

And what am I afraid of behind that? And behind that?
And behind that?

At my core, what is *the* thing I am most afraid of?

What am I afraid will happen?

WHOLE

When have I felt whole? What makes me feel whole?

What does *whole* mean to me?

Is the experience I'm having right now leading me to wholeness or drawing me away from wholeness? How?

What is that information telling me? Where do I go from
here?

INNER/OUTER

Does my outer world reflect my inner world?

How well does my life resemble me—the me I think I am
or the me I want to be?

LOVE

What do I love? *(This question doesn't have to be and probably
shouldn't be a who question. Focus on the what—the essence, the
spirit that brings you real joy.)*

Where is this love in my life? Where do I run into it, experience it, feel it?

What needs to happen to have more of what I love in
my life?

✱

SEEK

What do I seek? *(Do not gloss over this question. It may well be the most profound you can ask. In the mesmerizing documentary* Into Great Silence, *the abbot of the Grand Chartreuse, an ascetic monastery in the French Alps, asks a novice approaching the altar for induction only one question, "What do you seek?" The man answers with a clear, firm voice, "Grace." Think about it. That is an answer that can only come from the deepest fathom of the soul. You don't have to enter the cloistered life to consider and reconsider this profound question many times in your life.)*

peace

POWER

To whom or what have I given my power?

Why did I give away my power? What did I gain? Am I still benefiting from giving away my power?

How can I gather it back?

What will the price be? Am I willing to pay it?

HEART'S DESIRE

What is my heart's true desire?

Do I have any of it? How much?

Where did I get it?

How much more do I want?

What needs to happen for me to get more?

To serve doing something I love

SECRET BELIEFS

What are my true, deep, secret beliefs about myself and what I deserve?

At my core, do I believe that I am deeply and profoundly
    loved? Do I believe that I am worthy of divine guid-
    ance and grace?

What is keeping me from believing I am unloved and un-
    worthy?

What needs to happen in order for me to truly believe that
    I am loved and worthy of divine guidance and grace?

How can I change my most secret beliefs? *(If you've been
    wondering why your life isn't working out the way you want,
    these are questions to ponder—and ponder deeply.)*

[handwritten: Shame / being female]

[handwritten: accept / self totally]

[handwritten: total / surrender]

[handwritten: FORGIVE / SELF]

[handwritten: forgive / self]

AFRAID TO ASK

What one question am I afraid to ask right now?

What one question do I need to ask right now? *(When your
    conversation is going nowhere, you're getting frustrated, and you
    feel a need to break through to another level, ask the Voice these
    ultra-powerful questions.)*

PURPOSE

I have a divine purpose, a reason*[handwritten: (s)]*for being. What is that
    purpose?

How can I discern what it is?

How can I learn more about it?

How can I be sure that I've identified my true purpose and
    not just made one up that sounds good or appeals to
    my ego?

How do I know when I am living in sync or out of sync
    with my purpose?

Dear Voice,
Do I want to really probe deep within my soul?
Or am I afraid I'll find those monsters of the deep?
Help me. Help me choose a question to begin
this exploration. Then I can ask it, knowing I am
guided and protected. Where shall we begin?

(4)

Category 4: Soul Questions That Support IMAGINING AND INCUBATION: When you begin to identify your soul's desires and purpose, you simultaneously feel the urge to start creating an external life that mirrors your internal one. The status quo will no longer do. As your soul shifts and grows, your current life starts to feel like a tighter and tighter jacket. You can't help starting to imagine a life that fits your soul's new expansion.

Writing down your soul is a spiritual practice, yes, but also a highly practical one. Its purpose is to connect you with the guidance, strength, and tools you need to create the life you want—the life you are here to live. In this category of questions you are your life's designer. Playing the "what if" game with the Voice, you throw ideas on the page, discuss them at length, and over time focus on the ones that resonate. As you incubate the possibilities, they coalesce into something clear, something believable, something good. These questions are your sandbox, your tool shed, your playroom, your opportunity to visualize what life could be. But don't ignore that old adage: measure twice, cut once. Before you jump in and start creating a life, take time with these questions to design it.

WILLING TO SEE

Am I willing to see a better reality for myself? Do I regularly allow myself to see the possibilities?

What is keeping me from being willing to see a better reality for myself, or what ideas, emotions, fears, and experiences are blocking me from allowing a new and more delightful reality into my life?

POSSIBILITIES

What other possibilities are there? What lies beyond what I have experienced, what I can see, and what I currently imagine?

Are these possibilities real? How can I tell if these possibilities are real or not?

Which ones do I like best? Why do I like those?

What is the gap between where I am and my possibilities?

How can I close the gap?

WANT

What do I want? What do I really, really, really want? *(This is* the *question.*

What would my life look like if I had what I really want?

Why do I want this and not something else?

What needs to happen to bring what I want to life?

ANYTHING

→ If I could do anything, what would it be?

If nothing were in the way and I could truly do what I say
I want, would I do it? Why? Why not?

POWER

What is power—real power?

When do I feel powerful? Powerless? What is the difference between the times when I feel powerful and the
times when I feel powerless?

How much real power do I have? Have I ever used it?
When?

What is the source of this real power?

Do I want real power? Or am I afraid of it? What keeps
me from embracing and using my real power?

What difference would it make in my life if I had some
very real power?

What needs to happen for me to have steady access to real
power?

ALREADY SOLVED

What would my life look like if my problems were healed,
my prayers answered, and my dreams came true?

How would I feel?

How different is that life from the one I'm in right now?

What is in the gap between where I am now and where I
want to be?

Do I really want all my problems solved? What parts of
me really want that life, and what parts aren't ready?

What needs to happen in order for my whole self to want that life?

OUTSIDE/INSIDE

How can I create a life on the outside that looks like the me on the inside? In what ways am I blocking that congruent life from coming into being?

What needs to happen for my outside to start matching my inside?

IF I LOVED ME AS THE VOICE LOVES ME

If I loved myself as the Voice loves me, what would my life look like?

What would I be feeling, doing, believing?

How different is that life from the life I have now?

BLOCKS

FEAR

What is blocking me from getting what I want? What ideas? What beliefs? What relationships? What emotions? What doubts? What fears? What situations? What resources?

Am I ready to start removing some of those blocks? How do I know if I'm ready to start removing some of those blocks?

What would happen if I removed some of them?

Which one would I move first? What would happen? Second? What would happen?

POWER TO CHOOSE

Do I believe that I have the power to create my reality, to
   choose good outcomes, to create my life? Do I really
   believe that? Really? How do I know I believe it?
Is it true? How do I know it's true?

EVIDENCE

What evidence do I have that I am protected and loved?
What evidence do I have that I am divinely guided?
Do I want more evidence? Need more? What kind of evi-
   dence do I need? Why do I think I need it?
How can I get more evidence?

BECOMING

What do I need to do to manifest what I say I want?
What do I need to be thinking? Believing? Saying? Doing?
What does a life of manifestation look like?

WRITE DOWN YOUR SOUL

Dear Voice,
Wow, what would my life look like if it matched
my soul's desires? Tell me. Show me. Please!
This is something I want to explore.
I do want to start incubating a beautiful life.
How do I begin? I mean, how do we begin?

Category 5: Soul Questions That Support CREATING AND MANIFESTING: In the end, theory is not enough; we want a richer, fuller, more joyful life—and not in some distant future. We want it right here and right now. This last category of soul questions helps you focus on the practical, realistic steps that create that life.

But don't rush to these questions too soon. Taking action works only when it is based on a deep understanding of your heart's desires and a profound conviction in your soul's purpose. Without this foundation, you may toss a few bricks together, and things may even look pretty good at first, but then you'll get wildly frustrated when your building crumbles at the first tremor.

Don't look for a magic formula inside these questions. There is no one perfect process that works exactly the same way for everyone. What is here is an opportunity to build a beautiful life under the guidance and direction of the Master Architect.

WHAT IS NEEDED

What is needed right now? *(Store this question on the inside cover of your journal. When you don't know what else to ask, ask this. It applies to all times and all situations.)*

What question should I be asking?

HOW CAN I

I have a picture in my mind of what I want. I see it. Please help me build it. How can I turn this picture into reality? Who and what can help me?

When do I begin? How long will it take?

What needs to happen for me (I mean us!) to get started?

THINKING DIFFERENTLY

What do I need to change in my thinking to become the change I want?

How can I sustain that change in thinking?

How will I know when I've reverted?

How will I know when I'm expanding, moving forward, making progress?

SMALL CHANGE

What small change can I make right now?

What impact will it have?

Am I ready, willing, and able to make that change? What do I need to do to become ready, willing, and able to make that change? How do I know that I'm ready, willing, and able to make that change?

What needs to happen for me to implement that change?

LETTING GO

What do I need to let go of right now? What do I need to release?

What am I finished with? What is finished with me?

What will happen when I let go?

Am I afraid to let go? Why am I afraid? What's holding me back?

TRUST

Do I trust that I am divinely guided and protected?

How do I know when I am in a state of trust? What does trust look like? Feel like?

How do I know when I'm not in a place of trust? What
  does that place look like? Feel like?

What is trust? Where does it come from?

What needs to happen to fill my trust bank?

SURRENDER

How can I surrender when I don't know what will happen?

How can I let go when I'm afraid to let go?

Where's the safety net? Is there a safety net? How can I
  recognize the safety net?

THE GAP

What is in the gap between here and there—what I have
  and what I want, who I am and who I want to be?

How strong are these things in the gap? What power do
  they have over me?

How am I going to get from here to there? How can I
  cross that gap?

ONE THING

What one thing do I need to stop doing or start doing
  today?

What will happen when I make that change?

How can I sustain it?

What will let me know if I'm on track or off?

What if I change my mind?

FORGIVE MYSELF

Do I need to forgive myself? For what?

Why do I want to/need to forgive myself?

Am I willing to forgive myself? Am I ready? How do I know I'm ready?

How am I going to forgive myself?

How will I know I have truly forgiven myself? What clues will there be?

FORGIVE ANOTHER

Do I need to forgive someone else? For what?

Why is forgiving this person important? What difference will it make?

How is not forgiving impacting me?

Am I willing and ready to forgive? How do I know that I'm willing and ready to forgive?

How am I going to do it?

How will I know that I have truly forgiven?

WRITE DOWN YOUR SOUL

Dear Voice,
OK, I'm ready to start moving from where I am to where I want to be. Guide me. What needs to happen now?

Please don't treat these questions as a master list that you work through and check off. These are only a few examples of soul-probing questions. These categories and examples are offered to help you understand how to formulate questions that activate a cosmic response. As you work with these questions, your confidence in your ability to ask specific, real, meaningful questions will build. The strongest, most powerful questions are the ones that spring from you, your story, your desire to know.

Live the Questions

As you ask deeper and richer questions, you begin to see that the relationship of answers to questions is not a line; it's more of a box. The answers, you see, are *inside* the questions. When you ask the Voice a question, information starts flowing rapidly through your pen, and much of that information is (surprise!) more questions. Every time you get to a new place where you realize something, learn something, know something, you discover there's more to explore, more to uncover, more to learn. And so there is no end to your questions. These endless questions are a good thing. It's one of the beautiful mysteries of writing down your soul.

When you realize that there truly is no one final answer, just deeper and deeper exploration into the life you are here to live, you will stop searching for the golden answer. And here's the paradox: when you stop looking for *the* answer, you start getting answers—soul-stirring, eye-popping, life-changing answers.

Einstein said we can't solve a problem with the same level of thinking that created it. Well, pursue layer upon layer of soul-probing questions, and your thinking can't help but change. It's as if you put on a different head, different eyes, different ears, different mind. When you ask profound questions, you engage in an eagle-eye view of yourself from five miles up. You see information that was simply not available from your earthbound, me-bound, right-answer-bound perspective. Ask the Voice these kinds of questions repeatedly and your thinking—and with it your life—will change. But be patient. Allow the conversation to happen at your soul's own pace. The moment you turn your focus back to "This is *the* answer that will fix my life," that "perfect" answer will turn to dust and slip through your fingers.

Rainer Maria Rilke said this most eloquently in *Letters to a Young Poet*, which he wrote in 1929 to encourage a nineteen-year-old writer:

> I would like to beg you, dear sir, as well as I can, to have patience with everything unresolved in your heart and to try to love *the questions themselves* as if they were locked rooms or books written in a very foreign language.

||||||||||||||||||||||||||||||||||||||||||||||||||||||||||||||||||||||||||||||||||||||||||||||||||||||||||||||||||

WRITE DOWN YOUR SOUL

Dear Voice,
Help me understand what living the questions means.
What does my life look like when I'm living the
questions? What am I doing or not doing? How different
is living the questions from the life I'm living now?

||||||||||||||||||||||||||||||||||||||||||||||||||||||||||||||||||||||||||||||||||||||||||||||||||||||||||||||||||

Don't search for the answers, which could not be given to you now, because you would not be able to live them. And the point is to live everything. *Live* the questions now. Perhaps then, someday, far in the future, you will gradually, without even noticing it, live your way to the answer.

Capture Your Insights

As you ask and live the questions, you start connecting the dots between experiences in your life, your feelings about them, and what it all means. As you extrapolate meaning from your story, insights keep bursting onto the page. You get ideas you've never had before. You uncover repeating patterns. You begin to understand why you do the things you do. You get important insights into how your soul unfolds and grows.

You want to capture these rich nuggets as they bubble up from your soul. Don't think that just because they are important and meaningful to you at the moment that you'll remember them. You won't. As your writing deepens, rich new ideas will pour out rapidly and often, and your conscious mind can't—or won't want to—remember them. These insights, after all, are new. And all these new insights are going to create change, giving your internal critic and your old neural pathways fits in the process. So it's important to figure out a way to grab those insights and hold onto them for ongoing reflection and study.

Some writers find that talking with others helps them clarify and cement their new insights. If you typically "think out loud" or "brainstorm" this way, choose your partner carefully. You'll know you're speaking with a true friend of your soul if the conversation expands and enhances your insights and leaves you feeling stimulated and uplifted. On the other hand, if you have to justify your process, defend your ideas, or pull yourself together after the conversation, find a different partner or consider one of these other ways to capture your insights.

Some writers keep a small notepad nearby and jot their particularly rich insights in it. When they want to review those insights, it's easy to reread the small notepad rather than try to rifle through the big journal looking for particular passages. I know people for whom this technique works really well, but I don't want to stop to write in a second notebook.

During the years of my divorce, I'd slap a sticky note on the side of a journal page whenever I thought something profound had surfaced. My journals from those years look wild, with dozens of sticky notes of many sizes and colors poking helter-skelter out of the pages.

Recently, I came up with a better idea. I call it "Insights on the Left." I write my daily conversation with the Voice only on the right-hand pages of my journal. When I realize I'm hearing something to remember, I write it quickly in a few words on the wide-open left-hand page. When I'm finished writing, I glance at my left-hand pages and repeat my blessings and insights as I drink my water. Any time I want to review what I've learned in the past week, month, or year, I skim the left-hand pages. There are the distilled learnings of my soul.

Dear Voice,
I want to capture the insights that flow through writing
down my soul with you. How shall we do that?

Find Your Blessings

1   When you pick up your pen to write, you are unwrapping a present —several presents actually. But at first the only one you can see is a big battered cardboard box with your name on the label. This box is your story, and your story may be sad or difficult or lonely or stressful or painful. It may be a story of abuse, illness, abandonment, rejection, neglect, struggle, divorce, or death. As you open that box and explore your story, you find another box inside.

2   This second box, encased in jagged red metal, is your conflict. Every story, as any playwright, screenwriter, or novelist can tell you, has conflict. Take away the conflict, and you take away the lifeblood of the story. It is conflict that gives the characters the opportunity to rub up against life and against one another, and through the friction, learn and grow. Open the second box carefully and dissect what's inside. As you probe, you'll discover what your conflict really is, how it appears in your life, and how you respond to it. Somewhere in your conflict is a big key to your soul's journey.

3   Dig deep inside your conflict to find the third box. This one looks a bit more promising; at least it looks more like a present.

It's wrapped in your favorite color but the bow is intricately tied, and the ribbon is difficult to remove. As you wrestle the wrappings off, you see that this gift is your themes. The more you explore this present, the more you recognize your themes and how they appear over and over and over again at work, at home, and in all your relationships with people, things, money—even the Voice.

4   As you identify your themes, you discover a fourth box. This one is beautiful. It's wrapped in pale blue foil paper with a large delicate silver bow. Inside are all the things your soul is learning.

5   Deep within your learnings is your final gift. It is wrapped in gold, and it glows. The ribbon around this box isn't a ribbon at all—it's pure light. Inside are your blessings. This box contains the reason you're here, the reason you're digging, the reason you are writing down your soul.

Don't expect to unwrap each of these five boxes every time you write. In any one writing session, you may do little more than scratch at a bit of plastic tape. The first months of my writing practice, I only glared at the cardboard box, and upon realizing that it was addressed to me, screamed, "*No! No!* I don't want this! This is not the life I ordered! Take it away!" It took me a long time to just open the damn thing and start digging. It took me a few years of deep soul writing to reach my glowing gold blessings box and know that it was mine.

So be patient. You are where you are, and where you are is good. In fact, you can't be any other place. You can't rip the outer box open and shove everything frantically aside to grab that golden box. Why? Because your blessings are *inside* your learnings, and your learnings are *inside* your themes, and your themes are *inside* your conflicts, and your conflicts are in the one and only place they can be—*inside* your story. Trust the process and know that you are on the right path, and you will, in perfect time, touch that golden box. And when you do, you will look back at your soul's journey and say, "This is very, very good."

||||||||||||||||||||||||||||||||||||||||||||||||||||||||||||||||||||||||||||||||||||||||||||||||||||||||||

WRITE DOWN YOUR SOUL

Dear Voice,
What box am I exploring right now?
How do I feel about being at this stage in my soul's exploration? Am I content to be where I am?

||||||||||||||||||||||||||||||||||||||||||||||||||||||||||||||||||||||||||||||||||||||||||||||||||||||||||

Stop Writing When You Sense It's Time to Stop

No matter how powerful a writing experience is, at some point you have to put down your pen and step back into everyday life. Writing is a spiritual practice that supports life; it is not a substitute for life. So how do you know when to stop writing?

When Debbie Lane, an award-winning hypnotist, started writing, she experienced problems with time. From her first writing exercise, Debbie found she could slip effortlessly into theta and unconscious mind. Writing for Debbie was so deep and delightful that time simply evaporated. This experience was wonderful for her spiritual life, but not so great for her professional life. After she was late for two client appointments, she started setting a kitchen timer to zap her back into conscious mind.

People like Debbie have to watch the clock, but for the rest of us, it's more a matter of knowing. You experience a sensation of being heard. Sorry, I can't be more descriptive; it's a spiritual sensation not a physical sensation. Perhaps a bit of wisdom has come through. Perhaps answers are on the page or new and richer questions. Perhaps nothing has been resolved, but you still feel content, knowing help is on the way. A feeling of peace descends, and you simply "know" that the day's writing is complete. I sense I'm finished when an urge to write *amen* comes through my hand.

Allow yourself to follow that instinct and stop writing when the writing lets you know it's time to stop. This sensation will become clearer with practice. The key is not to stop too soon. If you find yourself repeatedly stopping before the dialogue is complete, you and your inner critic are probably afraid of the guidance you might receive and are cutting off the conversation before the Voice can respond. If you think you are stopping prematurely, set aside several long blocks of time for writing and ask the Voice high-vibration questions like:

What am I afraid of?

Why am I cutting you off?

Do I not trust you?

Why don't I trust you?

Or do I not trust myself?

Why don't I trust myself to hear you and know what to do?

What can I do to make this connection and hold it open?

How can I help myself be ready and willing to hear you speak?

How can I recognize when it's really time to stop writing?

Dear Voice,
Am I ending prematurely? Am I cutting you off?

Be Grateful

As you close, realize that you have received divine communication. This is something to be truly thankful for, so say thank you in whatever way is meaningful to you. When Lauralyn Bunn concluded my Akashic Record reading, she said, "I wish to thank the masters and teachers for their love and wisdom and guidance at this time." A similar sentence would make a powerful close to a writing session: "Thank you for your love and wisdom and guidance today."

I am so conditioned to write fast that even my close is swift. I write "TYG TYG TYG" across the bottom of the page. TYG stands for "Thank you, God. Thank you, God. Thank you, God." Sylvia, the woman who taught me to add water, simply writes "thank you." Other writers express gratitude with a quick flourish of the Möbius strip that illustrates writing down your soul. It's their way of saying, "I spoke to you, and you spoke to me." Experiment with your own unique way to express gratitude to the Voice. If nothing else, smile and nod your head.

Dear Voice,
I want to express gratefulness for this connection
and the wisdom and guidance I receive. How can I say
thank you in a way that's meaningful and personal—
something special between us?

Sign Off

Before you close your journal, sign your name. I know this sounds silly; the Voice knows perfectly well who you are. But remember how in step one, "Show Up," you wrote a salutation as if you were writing a letter: "Dear _____"? Well, in the same spirit, close with a signature, a mark, something that indicates that your personal communication with the cosmos is now complete. I close with a capital *J*. Donna Vernon signs off with "Your beloved daughter, Donna." Other writers close with a symbol they invent just for their soul writing.

WRITE DOWN YOUR SOUL

Dear Voice,
How shall we end our conversation? How shall I sign off?

Conclude Your Ritual

Look to your opening ritual for clues to your closing ritual. If you open with a prayer, for example, you might close with a prayer. If you raise your writing hand, you might raise it again. If you visualize the loving light of the universe, you might visualize it again coursing through you. If you touch your third eye, touch it again. If you are using water to deepen the experience, repeat the insights and blessings from your day's writing as you slowly and consciously drink the water. If you have a candle, blow it out.

||||||||||||||||||||||||||||||||||||||||||||||||||||||||||||||||||||||||||||||||||||||||||||||||||||||

WRITE DOWN YOUR SOUL

Dear Voice,
I like this idea of a closing ritual that
honors the experience of connecting with you.
What would such a ritual look like? What shall we do?

||||||||||||||||||||||||||||||||||||||||||||||||||||||||||||||||||||||||||||||||||||||||||||||||||||||

Then close your journal, put it in a safe place, and step into the rest of your day or night knowing you are blessed and loved. And know that the Voice awaits your next conversation tomorrow at the same time.

Concluding My Writing Practice

My statement of thanksgiving: _____

_____

_____

_____

My signature: _____

My closing writing ritual: _____

_____

_____

_____

Step Four: Follow Up

The purpose of writing down your soul is to receive guidance—focused, personal, specific guidance. You've done three things so far to attract that guidance. You've *shown up* to connect with the Voice, *opened up* and spoken from your heart, and *listened up* as your questions activated the Voice and the wisdom of the universe tumbled onto the page. Now what? What do you do with the guidance you've received? You can't exactly ignore it. Well, you can, but how silly. Why would you go through all this effort to ask the Voice for guidance, get it, and then turn around and say, "Hey, thanks, but I think I'll just keep doing things my way."

You are at this juncture for a reason. The Voice sent you an invitation and your soul responded. You pushed open that cosmic door and got a peek of the incredible feast the universe provides. Your soul sampled, smiled, and said, "More, please." And more is what you got: more questions, more ideas, more inspiration—and more guidance.

Now in *Follow Up*, the last step in writing down your soul, you have the blessed opportunity to bring that guidance to life. And guess what? You are the only one who can. The Voice can dish out all the guidance in the world, but your life doesn't start changing until *you* take action. But that brings up a few rather important questions. How do you know that your guidance is real? How do you know that it's the wisdom of the Voice on the page and not you or your inner critic blathering away? And even when you recognize the truth of the words, how do you know what to do? And what proof is there that this whole process of writing down your soul is actually making a difference?

Follow Up is all about answering those questions. I'll share many ways I and other deep soul writers recognize, verify, and implement our guidance. But please don't think these are the only things you can do. As your relationship deepens, you and the Voice will develop your own methods and a personal code and private symbols to go with it. The conversation, as you are about to discover, happens all the time, not just in the moments you write. As the masters and teachers of the Akashic Record said, "The potential to be connected is always there." That means the Voice isn't limited to the page and neither are you. There are all kinds of rich things you can do to keep that channel of communication open around the clock. Let me show you some of my favorites.

Recognize Your Guidance

Before you can feel comfortable taking any action, you need to feel confident that the guidance you are receiving is true. I asked Brian and Lisa Berman how they can tell when someone at a reconciliation workshop is experiencing a shift. "When someone connects with a deep inner truth," Brian said, "especially one that is opposite their conscious, expressed beliefs, tears well up." I often experience these kinds of tears at church. When the minister says something profoundly true, it seems half the people in the pews join me in the search for tissue. But tears of truth are not limited to spiritual settings. When you find yourself teary in response to a scene in a movie, a page in a novel, a song on the radio, or just an idea floating by in the air, your soul is signaling recognition.

Lisa Berman shared another way to recognize that a connection with truth is happening. "It's as if you have a balloon, and you put a needle in, and the air goes out and it's a relief: aaaaahh." This sensation happens to me, too. Sometimes, as I close my journal, I realize I've been holding my breath. I take a very deep breath and let it out. This release is always a clue that what has transpired is important, meaningful, and real. Or I find myself nodding. Unconscious nodding is another external sign of an internal knowing.

Sometimes, when I'm pouring my soul onto the page, full speed ahead, asking questions from deep within, I become so focused that the conversation and I become one. At these times, my eyes are no longer on the page. It would be more accurate to say that my eyes, especially my third eye, are *inside* the page, *inside* my mind, *inside* my soul. When I'm in that inside place, I can't tell you if I'm exploring my subconscious mind, the collective unconscious, the Akashic Record, or someplace else. I can't tell you if I'm in alpha or those glorious mystical theta brain waves. All I know for certain is that I am connected with *something* that is loving and wise— and that something moves my pen. The sensation can be fleeting, but oh so real. When this happens, the message that flows onto the page is profoundly true and exactly what I need to hear.

Sylvia became a committed deep soul writer as a result of a pen-moving experience two weeks after her first class. She emailed what happened:

> As I was writing, my pen sort of took over. "It" said I'm cutting myself short with the way I'm writing. It's been something to squeeze into my day, another item to check off my to-do list. My "pen" also told me that writing is actually something very gentle and loving that I can do for myself when and if I please. It's something I can immerse myself in, explore, examine, savor. It's a luxury that's free and yet invaluable, irreplaceable. It's the most enriching thing I could be doing right now. How could I want to rush that? To rush it would be a form of self-bullying, a way of saying I'm not worth it. It's saying, "Let me rush through what I know is important so I can get to what I've

accepted as being important." I'm so tired of living like that. I'm sharing this with you because I want you to know what writing is doing for me, and it's thrilling.

Recognition doesn't have to be as dramatic as pen movement. In fact, it doesn't have to be physical at all. Much of the time, it's a gentle feeling of resonance with something you know is true. It feels as if you are moving one step closer to your whole and holy self. No one else may see it, but you know it's happening and it is very, very good.

As you write, become aware of how your soul expresses that experience of recognizing and connecting with the truth. Maybe it is tears, releasing the breath, nodding, or insights you capture on the left side of your journal. Perhaps it is something else altogether. No matter what it looks like or feels like, in the end it is a *knowing*. As the masters and teachers of the Akashic Record said, "There is no question that you are connected. If someone said, 'Prove to me that you are connected with something,' you'd say, 'I don't have to. I know.'"

|||||||||||||||||||||||||||||||||||||||||||||||||||||||||||||||||||||||||||||||||||||||||||||||||||||||||||||||||||||

WRITE DOWN YOUR SOUL

Dear Voice,
How can I recognize my guidance? How do I know
that it's true for me? What clues do I get from
my mind, my heart, my body, and my soul?

|||||||||||||||||||||||||||||||||||||||||||||||||||||||||||||||||||||||||||||||||||||||||||||||||||||||||||||||||||||

But what if you don't know? What if you're confused? What if your guidance seems wacky, or scary, or too hard? What then? For starters, breathe. Know that you are safe and loved. Breathe some more.

Ask for Clarification or Confirmation

You can ask for clarification from the Voice in many, many ways. Start by asking high-velocity questions on the page. Questions like:

> I don't understand. What does this mean?
> I need to know more. How can I learn more?
> How can I know this came from you?
> Why is this information coming to me now? Why didn't I
> see it before?
> Am I hearing this clearly, seeing this clearly? What is it you
> want me to see?
> Is this true? Is this true for me? How can I know that this
> is true for me?
> If this is true, why does it seem so scary?

There is no end to the ways you can ask the Voice for clarification. And you needn't do it all at once. You can come back to an idea the next day or the next week or the next month. Keep writing until you feel firmly grounded in your guidance.

Another powerful source of clarification is dreams. Through dreams you get rich clues to what your soul knows and wants. When I'm wrestling with a problem, I often pose a question to the Voice as I'm falling into the theta brain-wave state that occurs naturally before sleep. Invariably, I wake with new insights into what is happening, what it means, and what I can do to move forward.

When Diana came home from her first writing class, she felt an urge to get started that very night. At the next class she told us what happened: "The words just poured out in an unstoppable stream. I found myself writing about my three marriages and asking over and over, 'What was *that* about? What was *that* about? What was *that* about? Tell me, I wrote, I want to know.' Finally, I fell asleep. That night I had a dream. In the dream I saw all three of my husbands, and in the morning, I woke with total clarity."

Not all dreams are quite so profound, but each one is a clue to what your soul is working on. If you want to become friends with this well of wisdom, ask for clarification before you sleep, then keep your journal or a notepad next to the bed and jot down everything you remember the next morning while you're still in the half-awake/half-asleep theta brain-wave state. Don't be surprised if you don't understand or are shocked by some of the images. Talk over your dreams with the Voice and together you can decide how to interpret them.

||||||||||||||||||||||||||||||||||||||||||||||||||||||||||||||||||||||||||||||||||||||||||||||||||||||||||||||||||||||||||||||

WRITE DOWN YOUR SOUL

Dear Voice,
Please help me understand my dream.
I asked for clarification on . . . and this is
what I saw in my dream . . . What is this telling me?
What do the symbols mean? What is my soul learning?

||||||||||||||||||||||||||||||||||||||||||||||||||||||||||||||||||||||||||||||||||||||||||||||||||||||||||||||||||||||||||||||

You can also ask for clarification in prayer. My favorite guidance-seeking prayer is "You shine the light, and I will follow." Before I do any following, I ask (well, OK, *demand*) that the Voice do some light-shining. I say this prayer a lot, but on the days when I need big guidance, I say it with real vigor. But beware! This is a dangerous, life-changing prayer. Because once the path is clear, you gotta take it. You promised. When the Voice holds up its end of the bargain, you have to start walking down that path. You are welcome to borrow my prayer, but I encourage you to find your own. You and the Voice can develop it together.

Often the light of clarification comes not in the form of a klieg beam on the road *to* take, but in the form of a closed door on the road *not* to take: You don't get the job. She moves away. He doesn't call back. They say no to your proposal. The bank turns you down. The deal falls through. You aren't chosen. All those things we call "bad luck" may actually be good guidance— and another way the Voice clarifies or confirms its wisdom on the page.

My son and I had a profound experience of closed-door guidance the fall of his senior year. Jerry sent an early decision application to a big-name university he desperately wanted to attend. On the last possible day, the fat envelope arrived. He was thrilled, but I was sick; the financial package was a disaster. I didn't see how I could possibly pay for it. But he was so proud of getting in, and he *so* wanted to go. I launched into my "You shine the light" prayer. I said it at least five times a day and I took my problem to the Voice. Two months later, the big-name school rescinded his admission. Jerry was devastated. I kept writing and praying. In the spring, he was admitted to a much smaller school that was more closely aligned with his learning style, and—here's the topper—he received a huge scholarship. When we walked around the campus for the first time, we both realized it was the perfect school.

Joy, the aesthetician with the broken heart, had a comical experience of closed-door guidance. After her fiancé walked out, she went into isolation. She spent four months sitting alone at home playing spider solitaire on her computer. Here's the odd thing: she won almost every game. Solitaire was the one little corner of her world where things went right. Joy also started writing down her soul and seeing a therapist. Four months later, she thanked me profusely for teaching her to write. "It's amazing how writing works. I realize now that this whole thing, as painful as it's been, is what had to happen. I'm actually glad it happened. I've learned things about myself I would never have learned if he had stayed. My therapist says I'm making really fast progress. But—and this is so weird—no matter how long I sit there, I can't win a single game of solitaire." I don't know how her angels managed *that*, but it sure looks like they gave Joy a place of solace when she needed it, then closed the door and locked the key when she was ready to move on.

You can also seek clarification or confirmation through card decks such as tarot, oracle, or angel decks. With these types of cards, you prayerfully ask a question and then choose a card, which gives you a response in the form of a symbolic picture or keywords. Using cards for clarification works because everything is energy—we are energy, the cards are energy—and all that energy is connected in the zero-point field. You, me, our journals, our divination tools—we are all connected. So our personal energy can resonate with the energetic intention of the author of a card set to give us the clarification we're seeking.

During the months of confusion about where my son was going to college, a friend gave me the *Goddess Guidance Oracle Cards* by

Doreen Virtue. Just looking at them, I was transported back to my nine-year-old self, who loved playing with holy cards. I blessed the goddess cards and put them right to work. But every time I asked, "What do I need to know right now to help my son," I drew the same card: Lakshmi, a Hindu goddess of abundance. The message on the card said, "Stop worrying. Everything is going to be fine." Well, from where I sat, everything did *not* look fine, but I always felt better hearing that someone out there in the universe knew that everything was going to be all right. After getting Lakshmi for the third time, I laughed, picked up my journal, and wrote, "OK. I get it. I hear you. Everything's going to be fine. Sorry I doubted. Thank you for taking care of us."

Some religious leaders frown upon using cards or other divination tools. They might recommend instead that people open their religion's sacred text and see what message God has for them. Whether you're randomly selecting a passage from a sacred text or randomly selecting a card from a deck, you're doing the same thing, because the energy that connects you to the sacred book is the same energy that connects you to the cards.

||||||||||||||||||||||||||||||||||||||||||||||||||||||||||||||||||||||||||||||||||||||||||||||||||||||||||||||||||||||||||||||||||||||

WRITE DOWN YOUR SOUL

Dear Voice,
I thought I wanted _____,
but no matter what I do, it isn't happening.
Is this bad luck or good guidance? How can I tell?

||||||||||||||||||||||||||||||||||||||||||||||||||||||||||||||||||||||||||||||||||||||||||||||||||||||||||||||||||||||||||||||||||||||

Books can be a powerful source of clarification. Have you noticed that once you hear about a book, you see it everywhere and everyone is talking about it? That's the universe nudging you. After three people told me I'd adore Elizabeth Gilbert's *Eat, Pray, Love*, I bought it. I'd been struggling with the scary thought that maybe I was the only person on the planet whose life was saved by soul writing—and if so, what was I doing writing a book about it? I desperately needed reassurance that the Voice was real and writing down your soul heals. On page 328 of Gilbert's book, I found it:

● "I love you, I will never leave you, I will always take care of you."

 Those were the first words I ever wrote in that private notebook of mine, which I would carry with me from that moment forth, turning back to it many times over the next two years, always asking for help—*and always finding it*, even when I was most deadly sad or afraid. And that notebook, steeped through with that promise of love, was quite simply the only reason I survived the next years of my life.

When I read that passage, I burst into tears. Elizabeth Gilbert was using the same writing technique I use to get in touch with her soul, and divine consciousness was responding to her the same way it does for me. It was exactly the clarification and reassurance I needed, and I whispered, "Thank you."

 Sylvia's preferred source of clarification is a brief reading in the workbook for *A Course in Miracles*. One summer, she was trying to figure out how to bring new revenue streams into the family business.

When I'm trying to solve a problem, I write my desire and then my positive and negative expectations. I realize that my negative expectations are powerful blocks. I need to see them on paper. I wrote my desire to diversify and then my positive and negative expectations. I was surprised by my negative expectations: My husband won't listen to me. He doesn't want to feel like I'm smarter than he is. I didn't come up with any answers, I just wrote them down. That morning the reading in *A Course in Miracles* was "I will receive whatever I request." And that's exactly what happened. I never said a word to him, but a few days later my husband said, "You know, Sylvia, we really need to diversify."

Sometimes books speak to you before you even open them. Have you walked through a bookstore or library, and a book fell into your hands? That's guidance. Have you looked at a cover and felt an overwhelming urge to read that book? That's guidance. Remember my sweet Great Dane puppy, dragging *The Artist's Way* down the hall? That was most definitely guidance. When a book calls to you, open it. It just may be the Voice tapping you on the shoulder and saying, "Here's a little wisdom for your soul."

Just as cards and books can clarify and expand our connection with the Voice, animals can be conduits for blessings and messages. At first, you may not recognize their messages, but if they cross your path often enough, eventually you will.

My animal messenger is an osprey. The day I woke knowing I was afraid of my husband, an osprey flew into the tree outside my bedroom window. After a few days, it dawned on me that the osprey might have a purpose. I looked up hawk in *Medicine Cards*, a card deck that teaches the power and wisdom of animals based on Native American traditions. Hawk, according to the cards, is a messenger from Spirit. I went straight to my journal. "Dear God, Thank you for my osprey. Tell me why you sent him." My guardian stayed there day and night, even in violent storms, clinging to a narrow branch hanging eight feet over the dock. Before I went to bed, I stepped onto the dock and said my prayers under the gaze of my guardian. Sometimes, I'd step forward a foot too far just to hear the glorious whoosh of his wings, but then I'd apologize, step back, and he'd return to his branch. During the day he'd screech at me from his perch on the mast of a sailboat behind the house. I'd run into the back yard and let him see that I was fine. Eighteen months later, I signed the real estate contract selling my house. That day, I went outside and took my usual position on the dock. "You've done a beautiful job protecting us," I said to my guardian between tears. "Thank you. But I can't stay here anymore. I have to sell this house. We are safe. You can go now." He spread his wings to their full five-and-a-half-foot span, took off, and did not return.

Emily, the young woman who abruptly left Chicago, told me about her recurrent snake dream. In her dream, everyone is afraid of a huge yellow snake, but she moves it gently aside so others can get by. Before I could tell her that snakes represent transformation, she said, "You know something? I've been writing about my snake dream and I think this snake is a symbol of how I'm changing. I don't want to see a snake in real life, but I like seeing it in my dream."

Whether they appear in person or in dreams, animals can be powerful symbols, delivering divine messages for our soul's important questions. When you want off-page clarification or confirmation about something you've been exploring with the Voice, try asking to see an answer in the form of an animal. If you already have a connection with an animal, ask to see that animal. If you don't have a relationship with an animal messenger, ask the Voice to help you. For help recognizing and understanding animal messengers, refer to the *Medicine Cards* or *Animal Speak* by Ted Andrews.

Once you've indicated that you are open to animal messengers, start paying attention. When your messenger appears, ask, "Why are you here?" That's what I did when I found a dead osprey on the side of the road. I had just purchased my new home and I'd been waking in the middle of the night with money sweats. I'd been asking the Voice for help taming my anxieties, and although I'd received some comforting messages, they just didn't seem to be taking hold of my soul. I kept asking for more help. Then, one day as I was driving home, I saw a dead osprey on the side of the road. I couldn't bear to leave such an exquisite bird to rot in the hot sun, so I took him home and laid him gently on the kitchen counter. I stared at him closely, marveling at his beauty and power. Suddenly I put my hand on his heart. "I know you're here for a reason. Tell me." Then, I grabbed my journal and had a rapid fire conversation with the Voice: "What is an osprey? A fish hawk. What is a hawk? A messenger. What is a fish? The symbol of Christ. What did Christ say more than anything else? Fear not." Within moments of picking up my pen, I had my message—and it was exactly what I needed.

||||||||||||||||||||||||||||||||||||||||||||||||||||||||||||||||||||||||||||||||||||||||||||||||||||||||||||||||||||||||

WRITE DOWN YOUR SOUL

Dear Voice,
I like the idea of having animal messengers. I think I'm
ready for that. Let's talk about it.
||||||||||||||||||||||||||||||||||||||||||||||||||||||||||||||||||||||||||||||||||||||||||||||||||||||||||||||||||||||||

There's another form of guidance in that story—synchronicity. A few weeks before I saw the osprey on the road, I'd heard a lesson at church about what Jesus said the most. When I asked the Voice what message the osprey had for me, I had no conscious recollection of that lesson. Would I have reached the same conclusion about my osprey's message if I had not heard that lesson? I don't know. But it does seem that so many experiences in life happen with timing that can only be described as divinely guided. As you write down your soul every day, you'll become more attuned to the many synchronicities in your life.

WRITE DOWN YOUR SOUL

Dear Voice,
Something odd just happened and I don't think
it's totally an accident. I think your hand is in this.
Here's what happened . . .

Once you become conscious of one type of divine assistance—be it dreams, cards, books, animals, or synchronicity—you may begin to notice other ways the universe guides and supports you. Maybe you pay more attention to music, art, nature, or conversations with friends. Maybe you begin to be more aware of the love and support of the angels and saints of your spiritual tradition or of family members who have passed on. Or maybe you begin to notice numbers that keep appearing in your life.

For me it's eleven. I declared my covenant on November 11. My osprey messenger arrived on February 16, 2000, a date that adds up to eleven. Throughout my divorce, it seemed that every time I looked at a clock it was eleven after the hour. Once I became conscious of elevens, they seemed to be everywhere. To me they are a reminder that I am safe and loved. Last spring, as I removed the cushion on the chaise lounge on my patio to clean it, a dime and a penny tinkled onto the bricks. I smiled and whispered, "Thank you."

Are numbers a form of clarification and confirmation for you? Not sure? Ask the Voice. For help recognizing the meaning of numbers, look at Dan Millman's *The Life You Were Born to Live.*

||||||||||||||||||||||||||||||||||||||||||||||||||||||||||||||||||||||||||||||||||||||||||||||||||||||||||||||||||||||||||||||||||||||||||||||||||||||||

WRITE DOWN YOUR SOUL

Dear Voice,
It seems that I see the number ___ a lot.
Is it a message from you? What does it mean?

||||||||||||||||||||||||||||||||||||||||||||||||||||||||||||||||||||||||||||||||||||||||||||||||||||||||||||||||||||||||||||||||||||||||||||||||||||||||

When you follow up on something the Voice has said by asking for clarification, that clarification can come in countless forms. How can you recognize what is your clarification and what is not? Pay attention. If something feels right, or at least intriguing, pursue it. If it doesn't, let it go. If you aren't comfortable with one way the universe appears to be communicating with you, ask for another. Just because a message or messenger speaks deeply and clearly to one soul, doesn't mean it's the right vehicle for you. The key is to notice and seek. Eventually you'll reach that indefinable but blessed place where you know—and you know that you know.

||||||||||||||||||||||||||||||||||||||||||||||||||||||||||||||||||||||||||||||||||||||||||||||||||||||||||||||||||||||||||||||||

WRITE DOWN YOUR SOUL

Dear Voice,
I sense that I am receiving messages.
How can I recognize my guidance?

||||||||||||||||||||||||||||||||||||||||||||||||||||||||||||||||||||||||||||||||||||||||||||||||||||||||||||||||||||||||||||||||

Deserve and Give Yourself Permission

All the blessings and guidance of the universe could be waiting for you as soon as you close your journal. The Voice could be lining up miracles, unlocking the door to abundance, arranging coincidences, and planting messages smack in the middle of your path, but unless you are open to seeing, hearing, and receiving, no amount of heavenly hand-waving is going to get your attention. How do you open yourself to all the Voice has to offer?

Start by giving yourself permission to walk down the road to the Voice's Big House of Receiving, climb the steps, and raise your hand to knock on the elegant, if slightly intimidating, door. As you stand there, looking at all that exquisite inlay, give yourself permission to want something more, to expect something more, to believe that you deserve something more. Give yourself permission to ask for and receive miracles. Giving yourself permission is really quite straightforward: Either you believe you deserve, or you don't. You feel either worthy or unworthy, loved or unloved, guided and protected or judged and punished. Either you feel it's OK to open the door, walk in, and receive the gifts of the universe, or you think you should hover outside waiting for someone in authority to give you permission—permission you somehow have to earn.

I Deserve

Here's a quick exercise to help you determine how ready you are to knock on that glorious door. Read each item on this "I deserve" list and check off the ones you deeply believe. It isn't good enough to acknowledge intellectually that a statement is true (we all know spiritually, theologically, and every other way that these things are true); you've got to know it in your gut. Pay attention to how your body reacts. If your heart beats a warm little *yes* and a smile sneaks onto your face, check the box. If your eyes squint ever so slightly, or your stomach squeezes, or in your mind you hear a sarcastic little, "Yeah, right," move on.

- ☐ I deserve to have a better life.
- ☐ I deserve to find my way to the life I want.
- ☐ I deserve to feel the guiding hand of Spirit.
- ☐ I deserve to ask for miracles.
- ☐ I deserve to receive miracles.
- ☐ I deserve to be happy.
- ☐ I deserve to live in a beautiful home.
- ☐ I deserve to be healthy.
- ☐ I deserve to be in a loving, committed, partnership relationship.
- ☐ I deserve to have loving, joyful, peaceful relationships with _____ (fill in the names of children, siblings, friends, or whomever).

- [ ] I deserve to have work that brings me joy, meaning, and prosperity.
- [ ] I deserve to have financial freedom and security.
- [ ] I deserve to receive the abundant blessings of the universe.
- [ ] I deserve to have a real, vibrant relationship with the cosmic divine.
- [ ] I deserve to be forgiven for my mistakes and transgressions.
- [ ] I deserve to live in peace.
- [ ] I deserve to be whole.
- [ ] I deserve to be healed.
- [ ] I deserve to be holy.

How did that go? Did you check most of them? If not, take a moment to study the ones you didn't check and ask yourself and the Voice why you think you don't deserve these good things.

‖‖

WRITE DOWN YOUR SOUL

Dear Voice,
OK, that "I deserve" exercise was interesting—and
awful. I'm not going one step further until we talk this
out. Why don't I believe I deserve? You gotta help me!
What do I need to look at? What do I need to learn?
What do I need to know? Show me.
Show me, because I want to start deserving.

‖‖

Do not skip over or rush through this deserving piece. Remember those crusty old neural pathways that hold you hostage to your old way of life? You just stumbled upon some really big ones. And, lucky you, now you know their names. So how do you change them? The answer lies *inside* the things you aren't quite ready to believe you deserve. Look at the items you did not check on the "I deserve" list. Select four or five. Don't take on too many. Start small and build up some spiritual muscle before you try to change everything in your world.

Write your four or five biggest bugaboos or the things you need the most help with right now in each of the numbered lines that start, "I deserve to. . . ." Under each "I deserve" statement, there are three columns. In the first, describe what you currently have. In the second, describe what you want. And in the third, write a sentence that describes what you want in the present tense as if you already received it. The whole chart is your "I Deserve" Plan.

To help you get started, I'll show you what my first plan looked like (see page 192).

Janet's "I Deserve" Plan

1. I deserve to have a peaceful relationship with my ex-husband.

What I Have	What I Want	What It Looks Like Fulfilled
I hate him and he hates me. I wish him ill, and he finds new ways to hurt me all the time. Our son hates visitation. This is a mess, and it has got to stop.	I want to be able to talk about our son, make decisions together, go to school events together. I want our focus to be on what's best for Jerry. I want Jerry to want to see his dad and have a good time when he's there. I want my ex to have enough money for child support, but I want him to stop attacking me in court.	My ex-husband is in his perfect place, and there is peace, harmony, and prosperity for all.

2. I deserve to live in a beautiful home.

What I Have	What I Want	What It Looks Like Fulfilled
My son and I are in a dark, ratty apartment furnished with worn-out furniture. The only "art" is cheap metal birds nailed to the wall over the couch.	I want to live in a home with lots of windows and real art on the walls. Everyone who enters our home comes in peace and love. I want to be happy in our house.	Jerry and I are in our perfect home, filled with beauty, love, light, and life.

3. I deserve to receive the abundant blessings of the universe.

What I Have	What I Want	What It Looks Like Fulfilled
I am constantly afraid about money. I'm worried I can't send my son to a good school. I'm worried about our future. I'm worried about paying the bills. As for retirement savings—it's a disaster.	I want the heavens to open and prosperity to pour down. I want to give my son a great education. I want to stop panicking about our future. I want to have money to invest. I want to make good decisions about money.	I am the beloved child of the all-loving, all-giving father-mother God. I am ready, willing, and worthy to receive my inheritance right here and right now. I graciously receive, joyfully spend, wisely invest, and properly share the generous gifts the universe bestows on me.

4. I deserve to have work that brings me joy, meaning, and prosperity.

What I Have	What I Want	What It Looks Like Fulfilled
I work alone in my ratty apartment. I'm lonely. I don't have enough work. I don't really like what I have. I don't see how anything I do really matters. I'm not making anywhere near enough money.	I want to make a good living doing something I love. I want to write. I want to make a difference through my writing. I want to work in partnership. I want to be able to give money to the people and places who have helped me.	I am doing my perfect work, joyous as a result of my efforts, and generating prosperity for myself, my family, my partners, my spiritual sources, and all the people we serve.

OK, now it's your turn. Here's the first item of a blank chart. You can photocopy it as many times as you need and fill in your answers on the copy, or use it as a template to recreate in your journal.

My "I Deserve" Plan

1. I deserve to

What I Have	What I Want	What It Looks Like Fulfilled

2. I deserve to

What I Have	What I Want	What It Looks Like Fulfilled

3. I deserve to

What I Have	What I Want	What It Looks Like Fulfilled

4. I deserve to

What I Have	What I Want	What It Looks Like Fulfilled

5. I deserve to		
What I Have	**What I Want**	**What It Looks Like Fulfilled**

Now, take the present-tense, "fulfilled" statements in the last columns and turn them into your "I Deserve Prayer Sandwich."

Start with a statement about the cosmic divine that moves you deeply. This statement is your "bread." Here's mine: "I live in the peace and strength of God knowing everything is according to divine will. I am safe and loved. Jerry is safe and loved." This bread statement reminds me that I am guided and protected and so is my son. It is true in every circumstance, and saying it never fails to comfort me.

Your bread statement could be anything from a statement of faith from your religious tradition (like that gorgeous Shema, "Hear, O Israel, the Lord your God is one") to a profound truth you read in a book, to a line in a poem, to something you developed through your conversations with the Voice. Don't get hung up on finding the *perfect* opening statement. Whatever you choose is a mirror of your current spiritual awareness; you can change it any time you like.

After your opening bread statement, write your four or five "fulfilled" sentences. They are the meat, cheese, veggies, and condiments, so to speak, in your prayer sandwich. Then close by writing your bread statement again.

It only takes a couple of minutes to "eat" your prayer sandwich. But, oh, is it powerful! Wrap your lips around your prayer sandwich two or five or ten times a day, and those crusty neural pathways have no choice but to collapse.

Mind you, you're going to feel like a complete fool the first few months saying your prayer sandwhich because, of course, none of these statements are true—not yet. But don't fall into the trap of saying them in future tense: "I *will* be safe and loved. We *will* live in our perfect home. I *will* be doing my perfect work." If you say them that way, when will they ever become true in the *now*? Remember, the unconscious mind doesn't distinguish past from present. To the unconscious mind, it is always now—because, as we learned in the Compassionate Listening Project and quantum physics, it *is* always now. So even though the fulfilled statements seem like complete lies, say your prayer sandwich as if they were all true right now, this very moment.

Here's the prayer sandwich I said through my dark time:

Janet's "I Deserve Prayer Sandwich"
Bread: I live in the peace and strength of God knowing everything is according to divine will. I am safe and loved. Jerry is safe and loved.
Turkey: My ex-husband is in his perfect place, and there is peace, harmony, and prosperity for all.

Lettuce: Jerry and I are in our perfect home, filled with
beauty, love, light, and life.

Tomato: I am the beloved child of the all-loving, all-giving
father-mother God. I am ready, willing, and worthy
to receive my inheritance right here and right now. I
graciously receive, joyfully spend, wisely invest, and
properly share the generous gifts the universe bestows
on me.

Mayonnaise: I am doing my perfect work, joyous as a result
of my efforts, and generating prosperity for myself,
my family, my partners, my spiritual sources, and all
the people we serve.

Bread: I live in the peace and strength of God knowing
everything is according to divine will. I am safe and
loved. Jerry is safe and loved.

I started saying my prayer sandwich in November 1996. I continually fiddled with the wording, but the basic ideas didn't change much over the next three years because I continued to need the same things: peace with my ex, a nice home, financial security, and joyful work. Each time I tweaked the words, I printed out the prayer and put copies on my desk and next to my bed. I said my prayer sandwich morning and night and every time I felt scared.

One morning, I woke up in my brand new townhome, light pouring in the not-yet-curtained windows. Out of habit, I went through my prayer sandwich in my mind while my eyes were still closed. When I got to the sentence about being "in our perfect home filled with beauty, love, light, and life," my eyes flew open and landed on my colorful Kay Carlson painting of sweet peas.

"Oh my God!" I gasped, "We *are* in our perfect home! It *is* filled with light. It *is* filled with beauty, love, and life." I dissolved into a prayer of thanksgiving.

All my original prayer statements eventually became real—even, or especially, the one about harmony with my ex. I still say my prayer sandwich every morning and every night, but now my meat and veggies are my son, my writing—and you.

It's your turn. Create your "I Deserve Prayer Sandwich." Add color, illustrations, fancy fonts—anything you like. Make copies and put them wherever you'll say your prayer: next to your bed, in your journal, in the car, where you work. Read it every morning and every night. Read it before you write down your soul or after you finish. Read it whenever and wherever you need to remind yourself of the beautiful life you and the Voice are creating.

My "I Deserve Prayer Sandwich"

Bread:

-
-
-
-
-

Bread:

I Give Myself Permission

Why is giving yourself permission important? After all, if it's blessings we're talking about; can't the Voice dish them out just fine without our permission? Well, yes, but the truth is, if you don't believe you deserve, your good can't come. It's like asking for something and holding your left hand out palm up to receive it, while holding your right arm straight out in front of you like an angry crossing guard blasting a whistle, "Stop right there!" Your good cannot get past that guard. The whole smorgasbord of life can be laid out in front of you, but if you can't put your "no" arm down, you can't receive a thing.

||||||||||||||||||||||||||||||||||||||||||||||||||||||||||||||||||||||||||||||||||||||||||||||||||||||||||||||||||||||||||||

WRITE DOWN YOUR SOUL

Dear Voice,
Do I have a stern crossing guard blocking me from receiving what I say I want? How do I know I have one?
How am I getting in my own way?

||||||||||||||||||||||||||||||||||||||||||||||||||||||||||||||||||||||||||||||||||||||||||||||||||||||||||||||||||||||||||||

How do you give yourself permission to receive what the Voice has to offer you? The following exercise is one way to get started. Read this list of permissions. Check the ones that make you want to pump your fist and shout, "Yes, I do!" Skip the ones that make your eyes go wide or your stomach jump. As you read, you may think of some new ones. Add them at the end.

- ☐ I give myself permission to discern what's good for me and what isn't.
- ☐ I give myself permission to let go of things, ideas, and people that no longer serve me.
- ☐ I give myself permission to think differently.

- [ ] I give myself permission to speak differently.
- [ ] I give myself permission to act differently.
- [ ] I give myself permission to see guidance everywhere.
- [ ] I give myself permission to recognize my guidance.
- [ ] I give myself permission to follow my guidance.
- [ ] I give myself permission to forgive myself.
- [ ] I give myself permission to forgive those who have hurt me.
- [ ] I give myself permission to ask forgiveness of those whom I have harmed.
- [ ] I give myself permission to want more.
- [ ] I give myself permission to ask for miracles.
- [ ] I give myself permission to expect miracles.
- [ ] I give myself permission to recognize my miracles.
- [ ] I give myself permission to receive and give thanks for my miracles.
- [ ] I give myself permission to feel worthy.
- [ ] I give myself permission to be happy.
- [ ] I give myself permission to say no to the things that don't make me happy.
- [ ] I give myself permission to trust that I am safe and loved.
- [ ] I give myself permission to love and be loved.
- [ ] I give myself permission to live in trust.
- [ ] I give myself permission to be in an authentic, committed love relationship.
- [ ] I give myself permission to know that I am the beloved child of an all-loving, all-giving creator.

- ☐ I give myself permission to live the life of a beloved child of a loving universe.
- ☐ I give myself permission to _____ _____.
- ☐ I give myself permission to _____ _____.
- ☐ I give myself permission to _____ _____.

How did that go? Did you mark all? Half? Some? A few? Pick the ones that feel especially good and write them on the left-hand column of a chart that looks like the one below in your journal. These are your rock-solid foundation. Whenever things are shaky or scary, reread them and know that you walk on solid ground.

Next, write the ones that make you a little woozy on the right, in the "not ready" column. Do you *want* to give yourself permission for these things? How can you *start* giving yourself permission? Don't worry if you don't have an answer.

I easily and fully give myself permission to:	I'm not ready to give myself permission to:

Here's a way of giving yourself permission—a way that has worked wonders for me and many others. Look yourself in the eyes in the mirror and say, "You are precious and important. I give you permission to. . . ." Finish the sentence with a specific permission you are struggling with or with an all-encompassing permission like, "I give you permission to live in joy," or "I give you permission to ask for and receive the blessings of the universe," or "I give you permission to know that all is well." If big statements like that are too overwhelming, start small with something like, "I give you permission to think differently."

Most people find this exercise excruciating. The first time I tried it, I flinched and shut my eyes. I had to say the words with my eyes closed for two weeks before I worked up the courage to whisper it while looking sideways at myself. It took three months before I could say them out loud, with vigor, while holding steady eye contact.

Here's my current permission statement:

✓ Janet's Permission Statements

You are precious and important.

I give you permission to be happy.

Now, write yours:

My Permission Statements

You are precious and important.

I give you permission to_____.

There's a huge side benefit to saying these words. "You are precious and important" is the most loving thing anyone can say. It takes the breath away. It packs more wallop than the often misused, "I love you." But we can't have the unconditional love we want until we first love ourselves unconditionally. Each time you whisper these words of great love to yourself, you are taking a sledgehammer to the old neural pathway labeled "not lovable enough" or "not good enough" or "not attractive enough" or "not smart enough." Everyone has a "not enough" hole. Whatever your "not enough" is, "you are precious and important" undoes it and replaces it with something way beyond good enough. Because the truth is, good enough isn't good enough. We want and deserve much more than that.

||||||||||||||||||||||||||||||||||||||||||||||||||||||||||||||||||||||||||||||||||||||||||||||||||||||||||||||||||

WRITE DOWN YOUR SOUL

Dear Voice,
What is my "not enough"?

||||||||||||||||||||||||||||||||||||||||||||||||||||||||||||||||||||||||||||||||||||||||||||||||||||||||||||||||||

Every time you pick up your pen to speak with the Voice or do any of the spiritual exercises in this book, you build spiritual muscle. But when you do this one in particular, you also expand your ability to love and be loved. Say "you are precious and important" until you know it in the core of your being, and someday you'll hear someone on earth say it, too.

~~Giving yourself permission to live a love-filled, joy-filled~~ life is a profound exercise. Don't skip it. As you give yourself fuller and richer permission, you'll find your dialogue with the Voice gets deeper and richer, too. And as your confidence in that exchange expands, you'll find your hands and heart spread wider and wider and you start asking for and receiving some very profound gifts—gifts we call "miracles."

Ask for Miracles

Thus far in your deep soul writing practice, you've learned how to speak from the heart, ask your questions, and receive and recognize guidance on the page. But can this divine dialogue go one step further? Can it elicit miracles?

Miracles are enigmas. We talk about them all the time, as in: "It's a miracle my daughter survived middle school." "The Cubs are gonna need a miracle this year." "It'll take a miracle to get through Thanksgiving without Uncle Harry getting smashed." For most of us, *miracle* is almost a comical word.

But *miracle* is also a power-packed word conveying the extraordinary acts of the divine, such as Sarah, wife of the Old Testament patriarch Abraham, getting pregnant when she was way too old; Moses parting the Red Sea; and Jesus feeding a crowd with a few bites of some kid's lunch. Those miracles were so big they became cornerstones of whole religions. We are comfortable with those miracles precisely because they are so old, but we don't actually think anything like that could happen now, to us.

The truth is, miracles are available to anyone, any time. They come in two forms. Miracle One is a view of life itself as a miracle. Albert Einstein expressed this approach elegantly when he said, "There are two ways to live your life, one as if nothing is a miracle; one as if everything is a miracle."

Think how beautiful life would be if each of us held the everything-is-a-miracle perspective all day, every day. The world would overflow with love and joy. I wish I could say that I always look at the world through Albert Einstein eyes, but I have to remind myself regularly that just being on this blue planet is a miracle. I have to remind myself that my body is a miracle, my friends are miracles, and my son is a miracle. Like everyone else, I slip all too easily into worry and fear.

||||||||||||||||||||||||||||||||||||||||||||||||||||||||||||||||||||||||||||||||||||||||||||||||||||||||||||||||||

WRITE DOWN YOUR SOUL

Dear Voice,
If I believed that everything is a miracle,
what would my life look like?

||||||||||||||||||||||||||||||||||||||||||||||||||||||||||||||||||||||||||||||||||||||||||||||||||||||||||||||||||

Worry and fear are the primary reasons why I regularly need Miracle Two—the right here, right now, make-it-better miracle. As you know from my story, I requested quite a few right-now miracles when my life was a wreck. And I always received them, although at the time I didn't understand what I was doing.

If you open a few sacred texts and read the accounts of the big miracles, you'll see that they follow a pretty standard formula: need, ask, know. First, someone is terrified. Second, the person turns to his or her divine source and says something like, "Look, I know this looks impossible, but I need this and I need it now, so I'm asking." And then comes the essential part, the miracle maker, so to speak. Once the request is made, the person *does nothing more*. This is key. Why does the asker stop with all the asking and doing? Because the asker *knows* it is being handled. Or, as a wise child in Sunday school told *Grace (Eventually)* author Anne Lamott, "You do what you can. Then you get out of the way, because you're not the one who does the work."

Smart kid. She's right—the divine does the heavy lifting. Your angels rig the computer for solitaire, or send the check, or arrange the coincidence. You just send out word that you need help and the universe picks up the vibration and responds.

We're really quite good at the first two parts of the formula: being scared and yelling for help. So how come so few of us experience miracles? Because of the third part in the formula—the *know* part. Jesus (who grasped a thing or two about miracles) explained it rather clearly: "Everything you ask and pray for, believe that you have it already, and it will be yours" (Mark 11:24). In other words *know* that your prayer is already answered. Know that the miracle is already here.

Eventually, you may become so attuned to the loving power of the divine that you can give up the need or being scared part. Take Daniel in the den. He didn't run around screaming. He *knew* Yahweh would hold the lion's mouths closed. Or take Jesus on the boat in a storm. Convinced the boat was going to capsize, the disciples woke him and begged him to *do* something. He looked at the wind and said rather calmly, "Be still." Once. He didn't repeat it. He didn't beg. He didn't tell the wind where to go or what to do. He just *knew* the storm would stop.

And, of course, if you know that you have your miracle already, you can't help but feel gratitude. That initial feeling of fear dissolves quickly and completely into the joy of thanksgiving.

But often when we ask for a miracle, we doubt severely that it is actually coming. We pester Spirit endlessly about how things are going. And we keep jumping up to do it ourselves. We tell the divine not only what we want but also exactly how it should look, when it should be delivered, and who should bring it. All in all, we have a distinct lack of knowing.

During my dark time, I became quite comfortable with the whole ask-know thing. I had no choice. But when life improved, I got out of the practice of asking. In May of 2007, I had a chance to revisit this profound spiritual practice. My problem was simple: I didn't have enough money to cover the month's bills. It had been a long time since I'd asked for a miracle, and I wasn't sure I still knew how to do it. Those dark-time miracles felt so far away that I wondered if they were even real.

I wrote to the Voice at length that morning about asking and receiving. I reviewed my previous miracles and remembered that knowing was the key. But that brought up a problem. What if I *don't* know? What do I do then? The Voice said trust anyway. The only way to reach the place of knowing is to let go of all the natural, normal, earth-bound disbelief and step into a state of trust. In other words, if you can't make it all the way to belief, at least suspend your disbelief.

"Do you believe that I love you?" the Voice asked. "Yes," I wrote. "Do you believe that I take care of you?" "Yes," I wrote. "Then trust that and the knowing will follow." As I wrote, my old friends, Trust and Know, slid softly back into my body. I was ready. I told the Voice what I needed and wrote, "I don't know how you're going to do it, but I know you are, because you always have, and you always do. I am safe and loved." I felt peaceful—not an intellectual appreciation of the quality of peace, but real peace inside my bones. I simply knew that all was well. I closed my journal, walked over to my desk, and clicked on my bank account to pay what bills I could. There, at the top of the screen, was an unexpected deposit of $1,200. I smiled and whispered, "Thank you." I was grateful, but not surprised.

Do you need a miracle? Are you ready to ask? Do you think you deserve to ask? Do you give yourself permission to ask? Do you trust that the divine loves you and takes care of you? Then ask. Ask knowing the Voice hears and responds. Ask trusting that your good is here—even if you can't see it yet. Ask knowing you have already received. Ask knowing you are a beloved child of a loving universe. Because that's what you are.

I learned a profound lesson on asking from my son when he was nine. It was a Thursday night, and he couldn't stop crying. The next day was the beginning of a long weekend with his dad and he just couldn't bear the thought of going. There was nothing I could do. I couldn't stop the visitation, and I couldn't lie to him and say that it would be all right. I wanted desperately to comfort him, so I tried something new. "Sweetheart," I said, "you can talk to your father's higher self." He sniffled and looked at me. I showed him how to make a triangle mudra under his heart. He jumped out of bed and put his hands together with the thumbs pointing up. "Now, from your higher self, call on your father's higher self and tell him how you feel and what you need." Jerry closed his eyes and spoke from a wise place deep in his heart: "Dad, I love you, but you know you can't take care of me. I don't want to come this weekend. Please don't pick me up tomorrow." He opened his eyes, kissed me, and crawled back into bed. He was asleep in minutes. A half hour later, his father called. "I'm not feeling well," he said, "will you keep Jerry this weekend?" My son told me years later that he called on his father's higher self five times, and four times his father cancelled visitation.

For children, this ask-know thing comes pretty naturally, but for adults, it takes time to master. We've had a lifetime to build up old neural pathways that tell us things like "I don't deserve," "miracles don't happen," "I have to earn everything I get," "life is hard," and "there isn't enough." If you have any of these recordings playing in your head, be patient with yourself. Start by noticing them. When they show up, ask the Voice to replace them with something true. Then write it. Write it several times until you believe it. "I deserve. Miracles happen. Miracles happen every day. The universe provides. Life is good. There is plenty for all." If believing these things is a struggle, ask for help. Ask for guidance. Talk it over with the Voice. Explore the parts of yourself that stubbornly hold on to deeply ingrained, self-defeating thinking patterns that hold you back from sampling the feast the universe has prepared for you.

WRITE DOWN YOUR SOUL

Dear Voice,
Do I believe in miracles? Do I believe that
I deserve to receive miracles? Or do I think
I have to earn everything? Help me figure this out.
It's important, because I want my life to get better.
What do I believe deep in my soul about miracles?

Karen, a single mom, leapt at the opportunity to ask for a miracle. For years, she'd told herself there wasn't enough. She bought things for her kids but never spent money on herself. When the subject of miracles came up in writing down your soul class, she decided to go for it. "I want sunglasses," she wrote, "and not those cheap, dollar-store glasses; I want the real thing." The next day she won a pair of $150 Ray Bans at a contest at work. "Wow," she said, "the Voice works fast!" Karen said this fun miracle gave her the impetus and confidence she needed to start exploring her issues with money. "This is going to be deep work," she said, "but I'm ready."

Now, you try. Ask for something. Ask for something you need. Ask with knowing. Ask, knowing you are always and forever loved. Ask, knowing the Voice wants nothing more than to give. Then do nothing. Trust that it is handled. No cheating. No peeking. No running into the kitchen to check on the divine cooks. If you feel you have to do something (which I totally understand, doing nothing is the hardest thing of all), tell the Voice, "I know you're working on this. Thank you in advance for taking care of me." Make this your new mantra. Add it to your prayer sandwich. Repeat it every time you feel that urge to jump in and manufacture the miracle all by yourself.

||||||||||||||||||||||||||||||||||||||||||||||||||||||||||||||||||||||||||||||||||||||||||||||||||||||||||||||||||||||||||||||||||||

WRITE DOWN YOUR SOUL

Dear Voice,
I am ready to ask for a miracle. I give it to you completely, no strings attached. I know that you provide this or something even better for the highest good of all involved. And I know you provide it in perfect time. Here is my request:

||||||||||||||||||||||||||||||||||||||||||||||||||||||||||||||||||||||||||||||||||||||||||||||||||||||||||||||||||||||||||||||||||||

Your miracle will appear. Maybe not in the form you expect or the timing you want, but it will appear. If the universe had given me the miracle of peace with my ex in the tiny form I could imagine, it would have looked like civility and a $432 monthly support check. Not bad, but nothing compared to the miracles I received. Thank goodness the Voice sees and gives more than we can imagine.

Keep the miracle channel open and you'll discover that Einstein was on to something: everything *is* a miracle.

Follow Your Guidance

In the end, writing down your soul is about accessing inner wisdom and receiving guidance. The final question—and a huge one—is, what do you do with that guidance?

Well, for starters, pay attention to it. Then take action. But hear this: just because the guidance is clear, it doesn't always mean that following it is easy.

In 2006, I received my first invitation to teach out of town. The timing was not perfect. I was in the midst of putting together a $350,000 proposal to develop all the Spiritual Geography materials. But I couldn't turn up a chance to teach at the beautiful Unity Center in Vero Beach, and I thought it would be fun to sleep in a funky hotel right on the Atlantic Ocean. Two hours into the workshop, I gave the group their first ten-minute writing prompt: "Dear Voice, Why am I in this class at this time in my life? Why is this important to me? What do I want to learn, discover, uncover?"

The eleven participants put their heads down and threw themselves into the question. I wanted to be a good role model, so I picked up my journal and asked the same question. Immediately, my hand wrote: "Do I have to write a book on deep soul writing?"

My eyes went wide. The last thing on earth I wanted to do was take on a big new writing assignment. There was no time, no energy, no money—no way! So I handled this question as elegantly as I handle all unwelcome guidance. I wrote, *"No! No! No! No!* I have a bank proposal to finish and you know it! I have a meeting next week with the banker. I have a poster to design, maps to order, a Web site to update. I'm working every day on Spiritual Geography. I don't have time to write a book. I can't do this! *No!"*

The guidance, however, was clear. When I got home, I sent a one-sentence email to Conari Press before I could talk myself out of it: "Would you be interested in a book proposal on deep soul writing?" Their one-word answer came the next morning: *"Yes!"*

I stared at the screen and mumbled, "Guess I'll be taking that guidance."

Debbie Lane, the award-winning hypnotist, also had a profound experience with following guidance she received from the Voice. She was scheduled to see a new and very challenging client with a debilitating medical condition that prevented him from working, going to a movie, or doing much of anything in public. Debbie felt a profound desire to help him, but there was no existing protocol for his condition. She picked up a pen and told the Voice: "I want to help Dave. I want to help Dave. I want to help Dave." She wrote for thirty minutes. When she finished, she had the outline of a brand new protocol. She used it in his first session. When Dave came out of trance, he looked across the room at Debbie and then down at his chest. He began to cry. For the first time in four years, his body was still. Now, whenever Debbie wants to know how to help others, she asks the Voice for guidance—and follows it.

Suzanne shared a sweet story about guidance with her writing down your soul group. She told us she was writing one morning about how difficult Valentine's Day was going to be for her mother because her father had died six months before. As Suzanne wrote, she received guidance to send her mother two roses with baby's breath and sign the card, "Love overflowing, Paul."

"Some people," she said, "might think this is a little freaky, but it felt right, so I did it."

Her mother called when the flowers arrived. "How did you know that he always signed his cards like that? Thank you. This is the best Valentine's gift ever."

Don't think that taking action means having to do something big. Most of the time you're guided to do something small—just what you need for the day or the moment. That's a good thing. The big stuff—move to Michigan, quit your job, end the relationship, go back to school, apologize to your brother, sell the house, stop fighting, sign the papers—can seem too big and too scary. If we only got big stuff on the page every day, we'd all stop writing. But the little guidance—the next piece of bread on the trail—is digestible and useful. As Rumi told God: "Nibble at me, don't gulp me down."

Jack was writing when a nibble came through: add pictures. Jack takes photos everywhere he goes, but until that guidance showed up on the page, he treated his writing and photography as separate activities. He started putting photos in his journal and discovered that they enriched his writing immensely. A photo he put in one day would often contain the seed of something he'd write about the next. After awhile, he started doodling and drawing, too. "Now my journal really is my soul's story," he told the class, "All of me is here."

Sylvia told me about one of her nibbles after an upsetting conversation with her college-age daughter.

I was really pretty miffed at her over her spending habits. I thought I had it all worked out as to what needed to happen. Then I sat down and started writing about it. As I wrote, I realized that yes, she absolutely needs to slow down on the spending, or she'll probably have to drop out. But I also realized that she's probably dealing with stress by spending. I then found the free counseling service Web page for her college. It sounds very good, and I will call them tomorrow. Something tells me that she will probably respond much better to some understanding and a proposal for help, than to ultimatums.

As Sylvia discovered, most guidance isn't about doing something; it's about shifting your thinking—and shifting your thinking can be tough. "What to do" guidance is really pretty easy to follow. I mean, there it is, on the page; what else are you going to do? When you see it—and your heart thumps in recognition—your feet automatically start moving, one in front of the other. But the guidance that tells you to shift your thoughts and feelings—that is so much harder to follow. Like that "love your enemy" thing that came through when I was livid with my ex. The guidance was unmistakable, but I could not bring myself to do it. I wanted to win, not change my thinking. I wanted to keep on hating him, not start loving him. But it's the change-your-heart/change-your-mind guidance that changes your life. It's what demolishes those old neural pathways and builds new ones. It's what leads to that new and improved movie Michelle Colt talked about.

Well, I sure wanted that new movie. I wanted a nice home, financial security, joyful work, and a happy child. It took me a while (OK, *quite* a while) but eventually I connected the dots that if I wanted those things, I had to change my mind. If I wanted a new life, I had to give up my old one. "All right," I finally told the Voice, "show me. Show me, and I'll do it."

Well, give an invitation like that to the Voice, and watch out! The guidance poured in every day. I gritted my teeth and followed it—often on my hands and knees. I quit talking about all the ways my ex hurt me and dissected instead all the ways I refused to see. I stopped fuming about his weapons and acknowledged my own. I stopped blaming him for everything and looked instead at how I created my mess. I stopped rehashing his betrayals and focused on my fixation with a happy-family fantasy. I stopped berating myself for my mistakes and forgave myself. I quit complaining about waiting for my ship to come in, and realized Spirit's been waiting a whole lot longer for me to wake up and start living the life I'm here to live. I stopped trying to control everything. This one was tough. When I pulled my thumb out of the "control" dike, it seemed my house, my son's education, my savings, and my business all went out with the tide. Every single day, I had to fight with myself to not revert to a blithering state of panic and blame. Every single day, I struggled to reframe my thoughts. Every single day, I forced myself to say my prayer sandwich and feel a shred of hope and joy. "Help me," I screamed. "Help me." And help came—on the page, every day.

After three years of following my daily guidance, I had shifted 180 degrees. What began as hate ended as loving forgiveness. What I had labeled a betrayal, I now viewed as a rich soul journey that brought me a new mind, an open heart, and an amazing relationship with the Voice. And, oh yeah—a new and much better movie.

The truth is, we create ourselves by what we choose to notice, feel, think, say, and do. The operative word is *choose*. Everything, as the masters and teachers of the Akashic Record remind us, is a free will choice. Maybe we don't choose *what* happens; but we do choose *how* we react. Bruce Lipton explains the power of the mind to consciously choose in *Biology of Belief*:

> [T]he mind is extremely powerful. It can observe any programmed behavior we are engaged in, evaluate the behavior and consciously decide to change the program. We can actively choose how to respond to most environmental signals and whether we even want to respond at all. The conscious mind's capacity to override the subconscious mind's pre-programmed behaviors is the foundation of free will.

Or as Ghandi more simply put it:

> Your beliefs become your thoughts. Your thoughts become your words.
> Your words become your actions. Your actions become your habits.
> Your habits become your values. Your values become your destiny.

We get to live the movie our choices produce. I had created an unhappy, pain-filled world. Following my guidance from the Voice, I shifted what I noticed, changed what I thought, adjusted my language, and altered my behavior. In the process, I transformed my destiny and created a much better movie. Could

I have made this new movie without the guidance of the Voice? I don't think so.

Recognize Your Critic's Masks

Don't think for one minute that your internal critic is going to sit still and let you create a whole new world. It doesn't want you to hear— never mind follow up on—any of the Voice's new directions. Why? Because then everything would change, and change, as far as your internal critic is concerned, is a very bad thing. So, if you want the Voice's guidance to come in loud and clear, you've got to get good at recognizing and countering your critic's many Halloween masks.

Its funniest mask is "Sleepy." It appears as the inability to wake up in time to write. To unmask your critic, put your soul journal beside the bed and hoist yourself upright the second the alarm goes off. You may actually find you love writing in those theta-soaked moments. If you have a problem with "Sleepy," ask the Voice: "What's with the sleep thing?"

Another comical disguise is "Oops"—as in, "Oops, my sister needs me," "oops, gotta help the kids with homework," "oops, there's the doorbell." Once you start recognizing your "oopsies" for what they really are, you'll laugh at your critic: Nice try. Very creative. Now, excuse me, I gotta write.

Then there's the "I Don't Know What to Write" mask. You have a rich soul filled with stories, questions, and desires, all bursting to be explored. If anything, you have too much to say, not too little. If your inner critic shows up in this mask, ask the Voice: "What is it I don't want to talk about? What is it I'm afraid to see?"

If none of those masks do the job, your critic will go for "Too Busy." This mask looks like chaos at home, drama in relationships, problems at work—in other words, life. But instead of encouraging you to throw your life on the page and ask for guidance, your critic will whimper: "See, you're too busy. You can't write. You don't have time." "Too Busy" can be a tough mask to counter because on the surface it looks like your critic is right; there isn't enough time. And the truth is, if you don't start figuring out why these dramas keep repeating, there never will be enough time. So get tough. Start pulling yourself away from other people's dramas. It's a sad truth, but most people don't want to change; they just want to vent. You, however, are interested in exploring your soul and moving toward wholeness. To do that, you have to protect your time and your energy. Carve out fifteen minutes for your soul. This once, put yourself first.

If your critic can't keep you from writing, it will at least try to keep you from spouting the truth. It'll put on a rather attractive "Surface" mask. Under this guise, it will encourage you to ask easy questions and settle for light answers. If you find yourself finishing quickly without working up a spiritual sweat, your critic is veering you away from the deep stuff. When you catch yourself skimming the surface, stop yourself and write: "Wait, that's not exactly the whole story . . . ." "Wait, I have another question," "What I really need to know is . . ."

Your critic might put on a creepy "Eyes" mask. This mask appears as fear that someone will read your journals, or worse, learn all your secrets after you keel over. If you fall for this one, you'll never really pour your soul onto the page. For starters, tell your critic to cool it with the fear factor. If it'll make you feel better, make a security plan and stick to it. If you need a post-mortem plan, ask a friend to destroy your journals or put instructions in your will. Do whatever you must to deflate the power of "Eyes."

When your critic needs to pull out all the stops, it will put on the black cloak of "Doubt." Doubt is tough. Doubt eats at the bones of your soul with thoughts like "What if no one is listening? What if I am bamboozling myself into thinking writing works? What if this whole thing about changing your movie is nonsense?" If those weren't tough enough, there's a new form of doubt circulating around the Law of Attraction. It sounds like this: "If I write about what's bothering me, aren't I just giving attention to the very thing I don't want? And by giving it attention, won't I attract more of it?" The answer is yes—and no. Yes, if your intention is to wallow in your woes, whining, blaming, and complaining, you will definitely produce more of what you don't want. But if your intention is to understand, release, and heal your soul's blockages, the answer is no. Don't let the false mask of "Doubt" prevent you from looking directly into your wounds and beginning the work of healing them.

Doubt is like an insidious cloud of mosquitoes, making your life miserable and potentially delivering a soul-weakening disease with every bite. Trust is your best doubt-abatement program. My favorite trust prayer is the "Prayer for Protection" written by James Dillet Freeman in 1940 for the armed forces. Thanks to Col. James Aldrin of Apollo 11 and Col. James Irwin of Apollo 15, it blesses us from the moon. The general public heard this prayer when Robin Roberts, the *Good Morning America* anchor, shared it on national TV. Say this aloud and see if it fills you with a wave of trust. (Change *God* to *the Voice* or the name you use when you write.)

The light of God surrounds me
The love of God enfolds me
The power of God protects me
The presence of God watches over me
Wherever I am, God is

I know I'm safe

And all is well.

I like to end this prayer with "And all is well." I know this prayer works. This is the prayer that brought my son safely home when he was in the car as my ex-husband committed a road-rage crime. But if it isn't sufficiently trust-inducing for you, write your own prayer. It can be as simple as "I know I'm safe."

The mask your critic wears is not important. What matters is that you understand that your critic is trying to scare you into staying asleep. Writing down your soul wakes you up, and if you want a new life, the first thing you must do is wake up. Don't, as Rumi says, go back to sleep.

> The breeze at dawn has secrets to tell you.
> Don't go back to sleep.
> You must ask for what you really want.
> Don't go back to sleep.
> People are going back and forth across the doorsill
> where the two worlds touch.
> The door is round and open.
> Don't go back to sleep.

||||||||||||||||||||||||||||||||||||||||||||||||||||||||||||||||||||||||||||||||||||||||||||||||||||||||||||||||||||||||||||||||||

WRITE DOWN YOUR SOUL

Dear Voice,
What masks do my critics wear? How can I recognize them? How effective have they been at keeping me asleep? Do I want to wake up? How can I stay awake?

||||||||||||||||||||||||||||||||||||||||||||||||||||||||||||||||||||||||||||||||||||||||||||||||||||||||||||||||||||||||||||||||||

Partner with Your Inner Critical Voices

Have you ever asked yourself: Who exactly are my inner critics? What do they sound like? What are they saying? And why do they work so hard to control me? I did. I got sick of my inner abusers reminding me of everything that was wrong and poking holes in all my plans. The night I signed a nice new consulting contract, they worked overtime telling me all the ways I could fail. Angry and exhausted, I sat up and screamed, "Stop it, guys!" My hand shot to my mouth. My abusers all had male voices.

The next morning I sat down in my writing chair, determined to have it out. "OK," I wrote. "That's it. Come out and say everything you have to say. Let's get this over with." It wasn't pretty:

> You haven't worked for a year. What makes you think you can handle this contract? Your savings are gone. How did that happen if you're so smart? How come your ex keeps pushing your buttons? How come you keep documenting how he abuses his son, and no one listens? If you're so smart, how come the judge didn't think it was an emergency? Ever wonder why he's able to sue you and drag you to court so easily? How come he hasn't paid his debts? That precious marital settlement agreement you wanted so desperately hasn't made him do squat. What makes you think you can write? Then why don't you have an agent? Think you can get a regular job? Ha! Who'd hire you? You're fifty, unemployed, broke, and don't want to leave your son for hours in aftercare. Yeah sure, you'll get a great job.

It was awful, but at least I had corralled their taunts into one place. I took a deep breath and steeled myself for my next question: "Dear God, I know this isn't you. Who is this talking?" The answer came quickly. I discovered my ex-husband was living rent-free in my head and so was my father at his worst. Right alongside them was an old boss I hadn't thought about in years. I had names, but I kept going; I sensed there was more. I asked, "Who else is speaking? Who else doubts my ability? Come out, wherever you are. Speak clearly in the daylight, so I can see you, hear you, and answer you. If you doubt my ability, speak!"

Well, they spoke:

> *You can't do it.* Says who? *Says me.* Says who? *Me.* Who? *Me, the one who knows.* What do you know? *I know you are afraid. I know you worry. I know you question your ability. I know you wonder if you're good enough.*

I'd been writing for almost an hour when I had my breakthrough:

> You're not my ex-husband, but you have his voice. You're not my father, but you sound like him. You're not my old boss, but you shout like he did. Who else do you mimic? *Anyone. Ex-bosses, ex-lovers, ex-husbands, ex-fathers, ex-teachers, ex-priests. Anyone.* Whoa. Wait a minute. I know who this is. I recognize you. You are the unholy three: doubt, fear, and worry.

I was so glad I didn't stop writing when I realized my inner critical voices were male or even which males. If I'd stopped writing down my soul at either of those points, I'd never have uncovered my real saboteurs.

Dear Voice,
I know these critical voices aren't you. Who are they?

Once you figure out who your critics are, you have a choice. You can try to destroy them or try to establish peace. I went for destruction. I wrote long anguished prayers burying doubt, fear, and worry. I wrote their names on slips of paper and burned them. I wrote new prayer sandwich statements calling in their opposites: know, love, trust. With each effort, I felt my soul move a scooch closer to love and trust, but significant quantities of doubt, fear, and worry were still in there—and I knew it.

I was ready to try peace. But I didn't know how.

I turned to Brian and Lisa Berman for help. "Can Compassionate Listening techniques help us make peace with our inner bullies?" I asked.

"Yes," Brian said. "If we listen to someone who frightens us, a belligerent neighbor, for example, we can see that person outside of ourselves. As a compassionate listener, we can listen to him, find out what is really of value to him, and what is his positive intention. And when we connect with him on that level, his negative behaviors diminish. So it is with our inner critical voices. Lisa and I have developed some compassionate listening work to listen to our inner critical voice as if it were an outsider and build a partnership to achieve an objective."

This sounded perfect. I asked my favorite question, *"How* do you do that?"

"Ah," Brian said, "it's the how part that is challenging. The basic exercise is to externalize the voice. First, each person writes down in a journal what their critical voices say and highlights one that triggers them the most. Then they train a partner to speak like the critical voice so they can hear it out loud as they hear it inside. The person who has this inner critical voice goes into a centering place. This is where the compassionate listening work comes. The person prepares to listen for the positive intention of the critic who lives inside. Then we role play the critical voice, working with the person to set up a mutually empowering partnership to live the critic's positive intention."

Ooh, a partnership with my inner bully. "How do you do *that?"* I asked.

Lisa said, "It isn't that the voice goes away. It's that you have a new script, a new tape that you can refer back to energetically. You've made a new cellular structure around how you hear the critical voice. The more experience you have establishing a non-reactive response, the quicker you can get back into partnership mode. Through this work you realize the inner critical voice isn't trying to hurt you. It is a response to an early hurt, a strategy to prevent you from being hurt in the same way again. Once you recognize its positive intention to keep you safe, you can have compassion for yourself—all of yourself."

"Think of it," Brian said, "as you are sitting around with all the parts of yourself. One part might be a chairman of the board, another part a small scared child, another part a huge yelling parent. There's a different level of empowerment when one sees that all the parts are integral to the whole."

WRITE DOWN YOUR SOUL

Dear Voice,
What do my inner critics' voices sound like
in my head? Have I heard these voices before?
Where? When? Who do they sound like?

Help me capture everything my inner critics say.
I want to get it all on paper so I can see it
and heal it. What exactly do my inner critics
say to me? When do they say it? What am
I doing that motivates them to speak?

What do my critics say that upsets me the most?

If I listen hard and dig deep to find my critical
voice's good intention, what do I find?
What are these voices trying to get me to do or
keep me from doing? How are they trying to protect
me? Where did that come from? When did they start?

So my inner critics aren't some evil spirits to fend off; they are shadow pieces of me that sprang to life in response to early wounds. This explains why I could never quite shake doubt, fear, and worry. Listening to Brian and Lisa's critical voice process, I remembered a similar process Elizabeth Gilbert described in her book, *Eat, Pray, Love*. In her darkest hour, she demanded that every experience she

Interesting-

had ever had of sorrow, anger, and shame come forward. She relived each moment and said to them one by one, "Come into my heart now. You can rest here. It's safe now. It's over. I love you." It took hours, but when she finished, there was, at long last, peace. "Nothing was fighting in my mind anymore," she said.

WRITE DOWN YOUR SOUL

Dear Voice,
I'm tired of pretending my shadows aren't me.
They are. I'm ready to embrace them.
Show me how. How can I become whole?

Now that I have this awareness of my critics' good
intentions, how can I hold onto it? I get upset
when they attack me, and it's so easy to slip
back into shame and fear. How can we be partners?
How can we work together for my soul's benefit?
The next time they speak to me, what should I do?

Now when I hear my critics—and I still hear them—I acknowledge their positive intention to protect me from taking big risks and getting hurt. Hearing their concerns, I invite them to join me in the Prayer for Protection. Then, all of me feels safe and loved.

Notice Your Evidence

Before I started writing down my soul, I wasn't a very grateful person. Life looked to me like a game of grab and snatch, and what I couldn't grab, someone else would snatch. I wanted that big bucket of happiness, but my eyes were somehow always focused on what was missing, not what I had.

Now, wherever I look, I see something I'm grateful for. When I step outside in the morning, I'm grateful for the queen palm towering over the pergola and the gurgle of the fountain and the scent of basil in the herb box. When I walk through my living room, I smile at my rust orange walls and at the glimmer of my mother's crystal. When I read, I appreciate well-turned phrases and compelling stories. When I cook, I'm grateful for every sizzle, smell, and taste, and for my chopping companion—Miles Davis. And it's not just things; I whisper thank you for ideas, conversations, and parking spaces. And when I glance at my son's photograph—well, my cup runneth over.

When you first begin writing, you may find yourself peering pretty far down into your cup to find something to be grateful for. It isn't that joyful things aren't there. They're there. You just may not be aware of them. To increase your awareness, start keeping track of your "evidence." Evidence is anything that crosses your path that makes you think, hey, the Voice is responding.

When the judge ruled that my son and I couldn't move from Florida to Wisconsin to live with my family, I stomped onto the porch, threw my arms up, and screamed, "OK, you clearly don't want this family to move. I don't know why, but you don't. But I do presume you want this family to eat! So here's the deal. I will tithe ten percent of everything you send, but you've gotta start sending—and soon!" Two days later, the phone rang. "You don't know me," a woman said, "but our corporate counsel said I should contact you to fix our hiring problems." By the end of the week, I had a shiny new $17,000 consulting contract. By the end of the year, I'd earned $108,000 and bought a townhouse. Talk about evidence.

But evidence doesn't have to be dramatic. It can be as small as a book recommendation, a question you hear on the radio, a Web site you stumble upon. A woman wrote to tell me how one of my "Religion and Spirituality" columns became her evidence. Her daughter-in-law had taken out a restraining order against her son. The woman knew her daughter-in-law's action was a product of anger, not danger. She prayed that morning for guidance to help her daughter-in-law forgive her son and lift the order. After she prayed, she did an Internet search for "bon appétit," looking for a luncheon recipe. My column that day was about how to make a prayer sandwich and ended with "Bon appétit!" Thinking she'd found a recipe, she clicked on the column. It was exactly what she'd prayed for. She sent the column to her daughter-in-law, who called in tears an hour later to say she'd asked the court to rescind the order.

Once you start connecting the dots between your requests and the Voice's response, you'll notice evidence everywhere. It may be tangible—like a feather in your path, a butterfly flickering by, a penny on the front step, a bloom popping through the snow, or free tickets to a concert. Or it may be intangible, like a call from an old friend, an accidental meeting, a valuable introduction. Whatever it is, collect your evidence and put it on a shelf—your "Evidence Shelf." Every once in awhile, sift through your evidence and revel in how dearly you are guided and loved.

Dear Voice,
What evidence do I have that you
hear me and guide me?

Celebrate Your Soul's Unfoldment

Being guided and loved is something to celebrate. But how? If you are in a Writing Down Your Soul group, you can share your evidence and miracles with people who nod knowingly and applaud enthusiastically. But what if you are sitting alone in your writing chair? Well, for starters, feel grateful. You are receiving guidance from heaven and just as you feel tickled when you receive earthly gifts, let that delicious feeling of gratitude swell in your heart for your Voicely gifts.

Rumi, always the startling poet of the human-divine connection, captures this rich idea in a few words: "If God said, 'Rumi, pay homage to everything that has helped you enter my arms,' there would not be one experience of my life, not one thought, not one feeling, not any act, I would not bow to."

Gratitude for the journey—the whole journey—changes everything. It's a filter through which you see your world differently. Brother David Steindl-Rast, who is equal part Benedictine monk and Zen student, is a master of the rich spiritual practice of gratitude and the blessings it confers. In his book *Gratefulness, the Heart of Prayer*, he writes:

> To bless whatever is, and for no other reason but simply
> because it is—that is our raison d'être; that is what we are
> made for as human beings. This singular command is en-
> graved in our heart. Whether we understand this or not
> matters little. Whether we agree or disagree makes no dif-
> ference. And in our heart of heart we know it.

||||||||||||||||||||||||||||||||||||||||||||||||||||||||||||||||||||||||||||||||||||||||||||||||||

WRITE DOWN YOUR SOUL *SOUL DAY*

Dear Voice,
What am I grateful for? What is gratitude anyway?
What does gratitude look or feel or sound like?
Am I a grateful person? How can I be more grateful?

||||||||||||||||||||||||||||||||||||||||||||||||||||||||||||||||||||||||||||||||||||||||||||||||||

By now, I'm sure your heart swells with gratitude for the guidance and wisdom that come to you through the practice of writing
down your soul. But sometimes you want to do something special
to express your appreciation. I know I do. Sometimes I want to do
something really big to celebrate and intensify my connection with
the Voice. That's when I set aside a Soul Day. There are no rules
about Soul Days; each person's day is a reflection of his or her
individual soul. But here are some ideas to get you thinking about
your own Soul Day.

SOUL DAY

Set your intention.

- The night before, ask for guidance for your Soul Day and put your journal beside your bed. Record your waking thoughts and dreams as you gently rise through theta waves. (Don't set an alarm clock the night before.)
- Look in the mirror first thing and say, "You are precious and important. I give you permission to have a glorious Soul Day."
- Declare your purpose: what do you want to learn, prepare for, request, or celebrate?

Pray and meditate.

- Refresh your traditions or try a new form of prayer, such as saying the ninety-nine Muslim names of God, singing a Gregorian chant, reading the Psalms, or praying outside to the four directions as Native Americans do.
- Pray with prayer tools you have not used before, such as prayer beads or prayer shawls.
- Listen to a guided meditation recording. (Visit the Virtual Spa at *www.serenitypathways.com* for breathing and meditation experiences.)
- Walk a virtual labyrinth at *www.gratefulness.org* or a real one if there's one in your area. Or use a finger labyrinth (available at *www.serenitypathways.com*).
- Have long conversations with the Voice, engaging all five senses.

Feed your soul.

- Read something that stimulates your soul.
- Listen to gentle, uplifting music (or enjoy silence all day).

- Sit in the sun, beside water, or in the woods.
- If you watch a movie, choose one that is gentle, loving, and profound.

Release your woes.
- Build a wailing wall and put your concerns in the cracks.
- Ceremonially release old fears and worries: burn them in a bowl, set them adrift in a river, or bury them.

Review your completed journals.
- Reread and remember your insights and special conversations.
- If you want to, destroy old journals by shredding or burning them, thanking each one for its contribution to your journey to wholeness.

Renew your partnership with your inner critical voices.
- Remember their positive intentions and thank them for their loving concern.
- Explore how they have helped you since your last Soul Day.
- Rededicate yourself to your partnership with them.

Renew and refresh your relationships.
- Call the spirits of the people in your life, whether living or passed, into your soul, meditate on the blessings you receive from them, thank them for being in your life, send each one grace and love.
- Forgive: finally, totally and completely forgive yourself and anyone who appears to have harmed you. Ask forgiveness to flow from those you have harmed.

Celebrate your blessings.
- Walk around your home and really look at everything you have. Bless each object and say a prayer of thanksgiving.
- Review your Evidence Shelf; feel gratitude for each gift.

Ask for guidance and clarification.

- Ask for clarification from cards or other divination tools.
- Ask for clear messages and messengers, then go for a long walk, paying attention to everything on your path.
- Give yourself the gift of an Akashic Record reading or other experience your soul is longing for.
- Discuss your guidance and messages with the Voice.

Ask and know.

- Identify your soul's desires for the next few months or year.
- Update your Prayer Sandwich, making sure to put those requests in the present tense.
- Make a vision board or other physical representation of your requests and post it where your subconscious mind can see it every day.
- Say a prayer of thanksgiving for your requests as if they are already received (because in divine time, they already are).

Express your soul.

- Paint, draw, or make a mandala. You needn't be an artist. Watch Deborah Koff-Chapin make a soul painting at *www.soulcards.com* to see how easy yet profound it is, then start playing with paints, colored pencils, or chalk. For insights into mandalas, check out *www.karepossick.com*.
- Dance or play an instrument.

Bless your bodymind.

- Do gentle body-mind movement like Pilates, yoga, tai chi, or qi gong.
- Get a massage, especially if the therapist can come to your home.

Eat joyfully (or fast, if you prefer).

- Light candles, set a beautiful table, make a soul feast of foods you adore.
- If you want to explore creating a holy, meaningful meal, read about the Jewish Seder celebration and create a personal Seder for yourself, right down to leaving an empty chair for the Voice the way Jews leave one for Elijah in the traditional dinner.
- Say grace before you eat. (If you don't have it, get a copy of *A Grateful Heart: Daily Blessings for the Evening Meal from Buddha to the Beatles*, edited by M. J. Ryan, and choose a grace from it every night.)
- Toast your soul's amazing unfoldment.

End your day with a grateful heart.

- Write one all-encompassing statement of gratitude and put it on your mirror, your refrigerator, your closet door—all over the house. Turn it into a scrolling screensaver on your computer. (At the moment mine is "I am happy and grateful for my many successful partnerships.")
- Soak in a long, hot bath.
- Say a prayer of profound thanksgiving for all the blessings of your day.
- Set the intention to dream about the messages of the day; record your dreams upon waking.

By no means is this a complete list. The power isn't in any particular activity; it's in your intention to honor your soul's partnership with the Voice. Create the Soul Day that inspires *you*.

I always have a Soul Day on New Year's Day—well, not a whole day, more of a soul morning. I always begin my year by blessing the past one, reviewing evidence of the Voice's loving guidance, and clarifying my intentions for the next twelve months. On New Year's Day 2006, I wrote at length about my heart's desires. I took a long time to remember my soul's journey through the past ten years. As I visualized how my story was unfolding, I confirmed that I was ready for the next stage. I wrote: "I'm ready for my voice to be heard. I'm ready for my marketing and publishing partners to appear. Show me what to do and I will do it." Then I did a reading with the *Medicine Cards* deck asking for guidance for the next year; the last card in the reading was Raven, which means that something magical is about to happen. I wrote a new prayer sandwich, drew it as a mandala, and taped it to my bedroom door. I felt peaceful and happy. I knew, even in those first few hours of 2006, that everything I asked for was heading toward me.

The next morning I started my new year by clearing out my email inbox. I was clicking along, delete, delete, delete, when I came to an unread book-marketing newsletter. I'd deleted several others like it, but for some reason I skimmed this one. Toward the bottom, a line caught my eye: "I don't know if this applies to any of you, but if you write on spiritual topics, you might want to check out *www.religionandspirituality.com*." I clicked on the link to the Web site. Instantly, I knew *this* was my forum. I spent the rest of the morning constructing an email to the editor. With a prayer, I pushed "send" and left for lunch. When I returned, there was a voicemail. I not only had a column, but I also started the next Tuesday! From my Soul Day request for marketing and publishing partners to my first column was a grand total of five days. Six months later, Conari Press sent me a "let's get to know one another" email. Everything I requested that Soul Day has since come to me.

WRITE DOWN YOUR SOUL

Dear Voice,
Help me set up my Soul Day. What is my
intention for the day? What kind of
day would support and nourish my soul?

||||||||||||||||||||||||||||||||||||||||||||||||||||||||||||||||||||||||||||||||||||||||||||||||||||||||||

We conclude this chapter on a very happy note. As well we should, because dialogue with the Voice is a joyful, soul-lifting, soul-expanding adventure. It beckons only one response from our throats: THANK YOU!

Before We Close

HERE WE ARE, AT THE END OF THE BOOK. I feel I haven't told you enough about writing down your soul. Did I tell you it is the sweetest, simplest (and richest) spiritual practice? Did I tell you anyone and everyone can do it? Did I tell you that the Voice doesn't care what name you use; the Voice just wants to talk?

Did I tell you that it's safe to open your heart? That it's vital to listen with your spiritual ear? That it's a blessing to befriend those inner voices that haven't always been so kind? Did I tell you all that? I hope so.

Did I tell you that questions are the magic that activates the Voice? And that the Voice responds in equal measure to your desire to know? And did I tell you that no matter how deeply you probe, there's always more to discover in your soul's divine adventure? And that bit about living the questions—I hope I shared that because it's such a profound truth.

Did I tell you that guidance and wisdom await—not just on the page, but everywhere? When you enter into the mystery of divine dialogue, it seems the whole world starts talking to you!

This is important: Did I tell you that the Voice is always listening? Even on a dark night when it feels like no one is present, no one is responding, no one cares—the Voice *is* there. Trust me on this, and hold on.

And did I tell you about miracles? Ask and know—that's really all you have to remember. But maybe the biggest miracle is the conversation itself. It's certainly a mystery: the Voice is out there somewhere in the vast, uncharted field, and yet it's right there on the page on your lap in the very same moment. Amazing.

Oh, and did I tell you not to worry? You can't make a wrong turn. It's true. You can't. Your soul is unfolding in its own perfect way. And the Voice is right there beside you—your witness, your partner, your guide.

I hope I told you this: writing down your soul is a wonder-filled experience, worthy of proper celebration, or at least an occasional whoop of joy.

Did I tell you that your life will change? Because it will—indeed, it already has.

One last thing: did I tell you the conversation never ends? It's really important that I tell you this. The conversation begins the day you pick up a pen and write "Dear Voice," and then it continues and continues and continues, until a day when you can no longer hold a pen, and then it continues in another place and in another way. The truth is, the conversation never ends. Isn't that divine?

Resources

Recap of the Four Steps to Writing Down Your Soul

HOW DO YOU WRITE DOWN YOUR SOUL? Here's the whole story in a nutshell. As you scan these recaps of the activities in each of the four steps, identify what you do regularly and fairly easily and what is more of a struggle. This exercise will give you a nice picture of where you are and what you can do to support the beautiful unfoldment of your soul. Come back and visit this recap often.

Show Up	Open Up
Set a time and place.	Begin with what's bothering you or happening to you right now.
Sit down.	Write what comes.
Write at your chosen time and place every day.	Explore your deepest thoughts and feelings.
Get ready to write: create and use your personal writing ritual.	Tell the truth.
	Write fast and ignore all writing rules.
	Tell your story.
	Speak from the heart.
	Break the silence.
	Begin to open your spiritual ear.
	Find the best method to access unconscious mind.
	Honor your soul's need for expression.

Listen Up	Follow Up
Listen as the Voice listens.	Recognize your guidance.
Be patient.	Ask for clarification and confirmation.
Create space for the Voice.	Deserve and give yourself permission.
Consciously want to hear the Voice.	Ask for miracles.
Ask for an understanding heart.	Follow your guidance.
Trust that you are safe and loved.	Recognize your critic's masks.
Be present in the now.	Partner with your inner critical voices.
Be willing to explore beyond conscious mind.	Notice your evidence.
Ask questions.	Celebrate your soul's unfoldment.
Live the questions.	
Capture your insights.	
Find your blessings.	
Stop writing when you sense it's time to stop.	
Be grateful.	
Sign off.	
Conclude your ritual.	

Books, Glorious Books

My Favorite Books on Writing

Baldwin, Christina, *Life's Companion,* Bantam, 1990

Cameron, Julia, *The Artist's Way,* Jeremy P. Tarcher/Putnam, 1992

DeSalvo, Louise, *Writing as a Way of Healing* (a thorough review of the writing-to-heal process), Beacon Press, 1999

Goldberg, Natalie, *Writing Down the Bones,* Shambhala, 1986 (not specifically about soul writing, but everything she says applies)

Lamott, Anne, *Bird by Bird,* Anchor Books, 1994 (not directly about soul writing, but wonderful, and laugh-out-loud funny)

Pennebaker, James W., *Writing to Heal,* New Harbinger Publications, 2004

Scientific Windows into What Happens When We Write

Hardt, James, *The Art of Smart Thinking*, Biocybernaut Press, 2007

Laszlo, Ervin, *Science and the Akashic Field*, Inner Traditions, 2004

Lipton, Bruce, *The Biology of Belief*, Mountain of Love/Elite Books, 2005

McTaggart, Lynne, *The Field*, Harper Perennial, 2002

————, *The Intention Experiment*, Free Press, 2007

Pennebaker, James W., *Opening Up: The Healing Power of Expressing Emotions*, Guilford Press, 1997

Pert, Candace, *Molecules of Emotion*, Scribner, 1997

Books that Celebrate the Spiritual Journey

Gilbert, Elizabeth, *Eat, Pray, Love*, Penguin, 2006

Hesse, Hermann, *Siddhartha*, Penguin, 1999

Lamott, Anne, *Grace (Eventually): Thoughts on Faith*, Riverhead Books, 2007

————, *Plan B: Further Thoughts on Faith*, Riverhead Books, 2005

————, *Traveling Mercies: Some Thoughts on Faith*, Pantheon Books, 1999

Rilke, Maria Rainer, *Letters to a Young Poet*, Dover Publications, 2002

Walsh, Neale Donald, *Conversations with God*, Putnam, 1995

Insights into Some Ancient Sacred Texts

Douglas-Klotz, Neil, *Prayers of the Cosmos*, HarperSanFrancisco, 1990

————, *Desert Wisdom: Sacred Middle Eastern Writings from the Goddess through the Sufis*, HarperSanFrancisco, 1995 (don't miss his Web site: *www.abwoon.com*)

Funk, Robert, Roy W. Hoover, and the Jesus Seminar, *The Five Gospels: What Did Jesus Really Say?*, HarperOne, 1996

Gottlieb, Lynn, *She Who Dwells Within*, HarperSanFrancisco, 1995

Nouwen, Henri, *The Way of the Heart*, HarperSanFrancisco, 1981 (a tiny book overflowing with the distilled wisdom of the desert fathers)

Gratefulness

Ryan, M. J., *A Grateful Heart*, Conari Press, 1994 (totally addictive)

Steindl-Rast, David, *Gratefulness: the Heart of Prayer*, Paulist Press, 1984

Insights into the Messages of Animals and Numbers

Andrews, Ted, *Animal-Speak*, Llewellyn Publications, 1993

Carson, David and Jamie Sams, *Medicine Cards: The Discovery of Power Through the Ways of Animals*, St. Martin's Press, 1999

Millman, Dan, *The Life You Were Born to Live*, HJ Kramer, 1993

Divine Poetry

Ladinsky, Daniel, *Love Poems from God*, Penguin Compass, 2002 (if you get just one book, let it be *Love Poems*)

———, *The Gift: Poems by Hafiz the Great Sufi Master*, Penguin Compass, 1999

Whyte, David, *The House of Belonging*, Many Rivers Press, 1997

My Favorite Card Decks

Carson, David and Jamie Sams, *Medicine Cards*, St. Martin's Press, 1999

Koff-Chapin, Deborah, *SoulCards*, The Center for Touch Drawing, 1995

Virtue, Doreen, *Goddess Guidance Oracle Cards*, Hay House, 2004

———, *Archangel Oracle Cards*, Hay House, 2004

Other Books Cited in this Book

Breathnach, Sarah Ban, *Simple Abundance*, Warner Books, 1995

Canfield, Jack, *The Success Principles*, HarperCollins, 2007

Emoto, Masaru, *The Hidden Messages in Water*, Beyond Words, 2004

Roth, Ron, *The Healing Path of Prayer*, Crown Publishers, 1997

People and Organizations

Brian and Lisa Berman, Berman Healing Arts
Compassionate Listening Training and Reconciliation Programs, theta
healing, sculpture
www.bermanhealingarts.com

Lauralyn Bunn
Akashic Record readings and practitioner training
www.akashicpathways.com

Dr. John Burton
Books: *Hypnotic Language: Its Structure and Use; Understanding Advanced
Hypnotic Language Patterns;* and *States of Equilibrium*
www.drjohnburton.com

Robert and Michelle Colt
Their consulting practice emphasizes consciousness and how to work with
the natural functions of the brain to release the pull of the subconscious
mind when it is in direct conflict with conscious desires
www.actingsuccessnow.com
www.insidegame.com
www.wholecoaching.com

Compassionate Listening Project
A nonprofit organization based near Seattle, Washington, teaching heart-
based skills for peace building and reconciliation in families, communities,
on the job, and in the world, even in the heat of conflict
www.compassionatelistening.org

Dr. James Hardt
Brain-wave training and research
www.biocybernaut.com

Debbie Lane, CHt
2007 Hypnotist of the Year, member of the International Association
of Counselors and Therapists and the Hypnosis Education Association,
recipient of the 2007 Hypnosis Achievement Award
www.wisdomhypnosis.com

National Guild of Hypnotists
www.ngh.net

Tom Nicoli, BCH, CI
Author, speaker, and certified instructor of hypnosis
www.tomnicoli.com

Differences Between
Writing Down Your Soul and Journaling

Not even I understood the depth and breadth of the differences between deep soul writing and journaling until I put them in this chart.

Writing Down Your Soul	Traditional Journaling
Begins with a personal writing ritual	Uses no particular ritual
Happens in a sacred writing space	Requires no specific writing place
Includes deep breathing	Doesn't mention breathing
Directly addresses the Voice	Entries are not addressed to anyone
Involves writing at the same time for the first thirty days	Requires no set writing schedule
May start with an agenda, but that agenda quickly flies out the window	Follows a goal or agenda
Is supported by scientific research	Doesn't mention scientific evidence regarding its effectiveness
Works best when you write fast	Uses a slow or normal pace of writing
Engages all five senses	Does not involve engaging all the senses
Requires you to not judge, evaluate or edit what appears on the page	Involves thoughtful consideration of what has been written and allows for editing, if desired
Assists you with exiting conscious mind to access subconscious mind and beyond	Provides no instruction on how to exit conscious mind
Enables you to enter the theta brain-wave state and reach mystical theta	Doesn't discuss accessing the theta brain-wave state or mystical theta
Activates the Voice with questions	May or may not utilize questions
Uses five productive types of questions and avoids four unproductive types	Does not mention specific question formats
Creates and maintains new neural pathways	May or may not create new neural pathways
Can be used to ask for miracles	Is not designed to ask for miracles
Provides a process for recognizing, countering, and partnering with inner critical voices	Provides no process for recognizing or addressing inner critical voices
Elicits guidance to improve your life	May or may not give the writer life-changing guidance

Thirty-Day Writing Log

Keep track of your first thirty days of writing down your soul on this simple log. Jot down the date and a few words that describe your writing experience for the day. At the end of the thirty days, you'll have a powerful record of the impact of your first conversations with the Voice. And you'll have created a rich new habit. I can see the old neural pathways crumbling from here!

Day	Date/time	Experiences, surprises, frustrations, insights
1.		
2.		
3.		
4.		
5.		
6.		
7.		

Day	Date/time	Experiences, surprises, frustrations, insights
8.		
9.		
10.		
11.		
12.		
13.		
14.		
15.		
16.		
17.		
18.		
19.		
20.		
21.		
22.		
23.		
24.		

Day	Date/time	Experiences, surprises, frustrations, insights
25.		
26.		
27.		
28.		
29.		
30.		

Permissions

The author gratefully acknowledges the following publishers for granting permission to quote passages of their books.

Passages from *Molecules of Emotion*
Reprinted with the permission of Scribner, an imprint of Simon & Schuster Adult Publishing Group, from *Molecules of Emotion* by Candace B. Pert. Copyright © 1997 by Candace B. Pert/Foreword copyright © 1997 by Deepak Chopra, M.D. All rights reserved.

Passages from *Writing to Heal*
Reprinted with permission by New Harbinger Publications, Inc. *Writing to Heal* by James W. Pennebaker, Ph.D. Copyright © 2004 by James W. Pennebaker, Ph.D. *www.newharbinger.com*

Passages from *Opening Up*
Reprinted with permission of Guilford Press, *Opening Up* by James W. Pennebaker, Ph.D. Copyright © 1997 by James W. Pennebaker, Ph.D.

Passages from *Goddess Guidance Oracle Cards*
Reprinted with permission of Hay House, Inc. *Goddess Guidance Oracle Cards* by Doreen Virtue Ph.D. Copyright © 2004 by Doreen Virtue Ph.D.

Passages from *Love Poems from God*
Reprinted with the permission of Daniel Ladinsky from the Penguin anthology, *Love Poems from God*, Copyright © 2002 by Daniel Ladinsky.

Passages from *Prayers of the Cosmos*
Reprinted by permission of Harper Collins Publishers, *Prayers of the Cos-*

Acknowledgments

My name on the cover makes me smile. The truth is there are many fingerprints on this book, beginning with my brother Larry in 1997. After reading my first deep soul writing exercises, he said, "You are writing something that does not exist in the literature. Keep writing." Thanks to Larry Conner, Spiritual Geography came to life. Thanks to Spiritual Geography, Larry Moffitt invited me to write for *religionandspirituality.com*. And thanks to those columns, Brenda Knight beckoned me into the Conari fold.

This may sound odd, but thank you to Jan Johnson for rejecting my original title. After three months, I turned to Stephanie Gunning, who came up with the perfect title in ten minutes. If that wasn't gift enough, my heart jumped when I heard Stephanie interview Robert and Michelle Colt. After my conversation with them, this book really took off, as the Colts introduced me to Dr. Hardt, who introduced me to Lauralyn Bunn, who introduced me to Ervin Laszlo, who ushered me into the sweet spot where science and spirituality speak the same language. And then there's my sister Mary Elizabeth, who out of the blue one morning said, "Hey, if you're interested in listening, you might want to look up the Compassionate Listening Project." A week later, I was in my first of several profound conversations with Brian and Lisa Berman, who carried the heart of this book to a whole new place.

Thank you to Rabbi Kate Fagan, who opened my eyes to an endless array of divine names, and Richard Hooper, whose knowledge of testaments—old, new, and forgotten—is breathtaking. Thank you to Tom Nicoli, who confirmed that change happens when you put it in writing, John Burton, who patiently answered questions about unconscious mind, and Debbie Lane, who let me experience the beauty and power of wisdom hypnosis. Thank you to Robin Saenger, who gave me *Love Poems from God* for my birthday, and Daniel Ladinsky, their inspired translator, who generously permitted me to share so many of them with you.

Thank you to Cheryl Harrison for her fearless vision and the creative ability to bring it to life. And a very special thank you to the intrepid spiritual explorers who dove into the adventure of deep soul writing long before it had any kind of worldly approval rating. Your stories and spirits are alive in these pages.

Thank you to the magnificent team at Conari Press, who truly are dedicated to producing "Books to Live By." Thanks to my editor Rachel Leach and copyeditor Amy Rost. They entered the pores of this book and made it sing. Thanks to Bonni Hamilton and Allyson May, who are with me every step of the way telling the world why they must—*simply must*—read this book.

Thank you to my sister Claire, brother Jay, and dear friend Pat; you carried me through my darkest time. Thank you to my precious son, Jerry. Everyone needs a wise and loving fan. I have mine. And Michael? Thank you. I am listening now. Last, thank you to the Big Voice, my ever-present, all-listening, all-loving guide. I hope my small lily voice has served you well.

Janet Conner, July 2, 2008
Ozona, Florida

About the Author

Janet Conner teaches workshops, retreats, teleseminars, and online courses on Writing Down Your Soul and other spiritual tools. She also teaches writers how to maximize their creativity. For more information on Janet, *Writing Down Your Soul*, and her upcoming events, visit *www.writingdownyoursoul.com*.

Janet created Spiritual Geography, the deep soul writing system that heals the broken heart. She wrote a column for UPI's *www.religionandspirituality.com* and now blogs at *http://janetconner.wordpress.com*. She lives and writes in Ozona, Florida. You can also contact her at:

Dear God & Company
P.O. Box 277
Ozona, Florida 34660
janet@writingdownyoursoul.com

To Our Readers

CONARI PRESS, an imprint of Red Wheel/Weiser, publishes books on topics ranging from spirituality, personal growth, and relationships to women's issues, parenting, and social issues. Our mission is to publish quality books that will make a difference in people's lives—how we feel about ourselves and how we relate to one another. We value integrity, compassion, and receptivity, both in the books we publish and in the way we do business.

Our readers are our most important resource, and we value your input, suggestions, and ideas about what you would like to see published. Please feel free to contact us, to request our latest book catalog, or to be added to our mailing list.

Conari Press
An imprint of Red Wheel/Weiser, LLC
500 Third Street, Suite 230
San Francisco, CA 94107
www.redwheelweiser.com